I LOVE JESUS BUT I HATE CHRISTMAS

TACKLING THE CHALLENGES OF BEING A CHURCH TECHNICAL ARTIST

W. TODD ELLIOTT

founder of the ***FILO Conference***

FILO.org

Design by Inkwell Creative

TABLE OF CONTENTS

ACKNOWLEDGEMENTS

Who knew it was so difficult to write a book? This has been a very long journey that started back in 2006. What started as simply writing down ideas in a document from time to time, then moved into writing a blog, has turned into a full-blown book. Why has it taken so long? Because. (Stop asking questions!)

In reality, it has required a huge support system for me to be disciplined enough to get this done. As the FILO Conference has grown from a small experiment in 2015 to multiple events in 2020, the FILO team has also grown and become a huge reason why we are finally crossing the finish line.

To the FILO Core Team, whether you've been directly involved in the book process, you have all helped to make this happen. Joanne, Dave, Aubrey, Cassie, Caleb, Delwin, Sarah, Nate, Jodi, Chelsea, words cannot express how much you all mean to me! A huge shout out to Chelsea who really provided the structure necessary for me to get my act in gear long enough to get to this point.

Thank you to all of my friends who helped read through this and give me such solid, constructive feedback to make the book as good as possible. I believe everything gets better with honest critique, and I so appreciate what each of you brought to help make the best resource possible for the FILO tribe.

Going back further in time with my thank yous, so much of what I learned about what I believe to my core, I learned from my time at Kensington Community Church and then had reinforced while I was on staff at Willow Creek Community Church. Mark, Steve, Kristin, thanks for taking a chance on someone without real experience and then teaching me how to become an adult and a leader. To then move to Willow Creek and be given a chance to lead in some pretty amazing situations was an experience of a lifetime.

To my parents for being such good examples of hard work and excellence. And for supporting me when I told you I didn't want to be an engineer, but I wanted to go work at a start-up church that met in a junior high school.

To my wife, Bissy, for being willing to go where we felt God was leading us at each turn. Working at a startup church in freezing cold Michigan. Moving to Chicago. Quitting my job. Starting a conference. You must be crazy! Thanks for loving me and supporting me through it all. I'm so glad I get to share life with you. To my kids whom I love so much. You didn't really help write the book, but thanks for being proud of your pops, poppi and/or buddy.

Finally to Ryan Morrill who came up with this book title back in 2002. Who knew it would become an actual book!

INTRODUCTION

Gear is easy. When I buy a new piece of equipment, it comes with an owner's manual. If I am trying to figure out what equipment to buy, I can usually find literature, a magazine article, or some other tech person who can help me figure out what I need.

What about dealing with the life I have chosen as a production person? A technical artist? Where can I go to learn how to be all that God has created me to be—a so-called techie? Where is the manual that helps me navigate the relationship with my music director? Or where can I learn how to deal with last minute changes that always seem to come my way?

As the title of this book states, I love Jesus, but why do I hate Christmas? For years, I wondered if I was doing something wrong. Was there something wrong with me? Have you ever had that thought cross your mind, yet you didn't dare express it because how could anyone hate Christmas?

It turns out, you're not alone.

Regardless of the size of the congregation or how big your budget is, there are challenges we all face in the tech world, and yet there are very few places to turn for help or answers. We all struggle with not having enough money, with church leadership that doesn't understand our world, with working too many hours at Christmas time ... OK, working too many hours all the time!

How can we navigate all these issues? Or are these struggles just the way it has to be? Since we all deal with them, it might seem like we have no choice. Yet, I believe we are called to change the world through the use of the technical arts in the local church, and that won't happen if we are constantly feeling victimized by the very churches we are called to serve.

I once led a discussion with about 20 technical artists from various churches near Calgary, Alberta, where we talked about issues technical ministries deal with. Halfway through the day, a couple of guys came up to me and said that everything we were talking about didn't apply to them. I'm sure I looked shocked.

They went on to tell me they served at a portable church and had many practical questions they wanted answered. It would have been crazy for them not to take advantage of a room full of other technical people from their city who could help them with their equipment issues.

When they finally asked their question, it was about whether or not to use electric drums. While this is a valid question, they weren't really asking about what equipment was right for their church. Their real issue was that their drummer played too loudly for the room, and no one wanted to talk to him about it. That has nothing to do with the difference between electric or acoustic drums. Many of the most difficult challenges we face as technical artists rarely involve gear but require more from us.

God has a special plan for you as a technical artist, and it's my prayer that what you read next will help you reach *your* God-given potential, so God's church can reach *its* full potential.

So, wherever you are on the spectrum between first-time volunteer and seasoned veteran, small church or mega church, I encourage you to open your mind and your heart to receive what follows. These truths apply to us all.

BOOK STRUCTURE

In an effort to make this resource as effective as possible, I have divided this book into four parts.

Part One will set up the idea of what it means to be a technical artist. I'll share my story and the journey I have been on over the years. We will talk about how it has shaped me, and how I think God designed the technical artist to function in the local church.

In **Part Two**, we'll talk about the practical, everyday production values that are key to success, regardless of the event or worship service.

Part Three builds on the foundation of production values and adds a layer of collaboration with other artists.

Part Four will look into how to effectively lead a group of technical artists.

At the end of each section, there are questions to help you discover who you are as a technical artist. Either as an individual or as a team, I would encourage you to let these questions stretch and challenge you.

As you make your way through this book, I hope the principles found in the following pages will help you become, not only the best technical person you can be, but simply the best version of yourself.

PART 1

TECHIE OR ARTIST?

1

IN THE BEGINNING

I began innocently enough. Like so many technically inclined people, I started getting interested in soundboards and microphones early in high school. The church I was attending with my family was just beginning to require sound reinforcement, and I learned as the church grew. I sat with a friend who ran the soundboard for our one microphone, and he trained me on the finer points of hitting the record button on the tape deck. It was big stuff.

Then one day it happened. The unthinkable.

My friend didn't show.

Now what?

Could I figure out how to run the sound board for the one mic AND operate the tape deck at the same time? As it turned out, I was able to manage the chaos.

From there, we added a few more mics, then a keyboard and then a (gasp!) guitar. These were heady days. It soon became obvious that I loved doing production work. I had no idea what I was in for, but the whole audio thing made sense to me. I became the kid at school who knew all about the sound system in the gym. I was the boy who fixed the warbled film projector. (If you don't know what I'm talking about, Google it!) When I ran sound at my church, things went well. When the music team saw me behind the soundboard, they knew everything was going to be all right. (Not that I didn't get complaints from some people who said it was too loud and could I please turn it down.)

One of the fortunate coincidences of my audio learning process was that my church's music director also led a performing group at my school that used lots of production equipment. I started out as a musician in this group but soon became the main tech for getting everything up and running for our performances. He and I developed a great relationship and an understanding of developing a great mix. This transferred to every event I mixed.

As time went on, one of the values that developed at our church was planting other churches. The associate pastor was charged with developing a team of people around him to plant a church, and he had a vision to start what he called a "Willow Creek" style church. None of us really knew what that meant, so while I was on a break from college in 1989, a group of us piled into a van and drove from Detroit to Chicago to attend a mid-week service at Willow Creek Community Church.

Before this, I was pretty clear on what production could do for a church service. You know: 16 channel sound board, 8 par cans, no gels; maybe drums, definitely not electric guitars; meeting in a high school gym. Even though I was knowledgeable, I was not ready for my first experience at Willow Creek. Not only was the building huge, but the service was like nothing I had ever been a part of.

There was a band. The band rocked! The mix was amazing. The lighting was inspiring. I didn't even know you could be creative with lights in a church. There were side screens with the lyrics on them. And not a hymnal in sight.

On the drive home, we talked about all the great possibilities for a church like this one in the Detroit area.

Shortly after this trip, Kensington Community Church (KCC) was formed in Troy, Michigan in 1990. I was still in college, but I was heavily involved working behind the soundboard and helping with setup when I was home on break. Right from the start, we were loaded with great production possibilities. We met in a middle school cafetorium—you know, a cafeteria with a stage. We used nasty folding chairs, and there was no air conditioning, just big noisy fans. Turning the fluorescent lights on or off was our lighting rig, and we hauled in our sound system from the back of a Ryder truck.

It wasn't much, but we were sold out to the vision of what God could do in the Detroit area through us. We did the best we could with what we had and made up for the rest with adrenaline.

After holding services in the middle school for about nine months, Kensington moved into the performing arts space at the local high school and then onto the brand-new high school's theater that sat 750 people. About this time, I graduated from Auburn University with a degree in Industrial Engineering and got a job working at Kensington.

During in my junior year, I decided not to go into engineering immediately but try to make a career out of mixing audio. I had no idea how that was going to pan out, but I felt like I needed to give it a shot. Fortunately for me, the people at Kensington knew when I was behind the soundboard things went better. My first job at KCC consisted of leading the audio team, writing out the music charts for the band and vocalists and leading the creative process for our mid-week service. It still makes me laugh to think about how I did all those things at the same time.

As KCC got bigger, my responsibilities focused more on the technical side, and I was able to stop charting music. It also became obvious that I needed to stop leading the creative process since I kept getting stuck on how to make it happen and constantly felt overwhelmed.

One of the things I learned early on in my years of working half the time in production and half the time in the creative arts was that I didn't feel like I fit into either category. I was a musician, but not a very good one. I could be creative, but only after someone else had come up with an initial idea. I was a technically minded person, but I didn't have the patience or the aptitude to learn more than what I needed in the moment. I was an Industrial Engineer but didn't necessarily remember all the details of any of my classes. I felt like one of those misfit toys from the "Rudolph the Red-Nosed Reindeer" movie.

Around 1995, KCC began discussing the construction of our own building. To help with this, we hired a consultant to guide us through the process of figuring out what we needed. One afternoon I told him I felt trapped between creative arts and production, and I didn't seem to fit into either category. He listened patiently, and then basically summed it up for me. "God has created you this way. You just need to be confident that God knows what he is doing and be patient that he will reveal to you what His will for your life is."

It's possible he was just tossing out advice to get me to shut up, yet it is something I have never forgotten. At the time it didn't hit me, but after about five years, it began to ring in my ears: "God made you this way. Be patient."

God has created you this way. You just need to be confident that God knows what he is doing and be patient that he will reveal to you what His will for your life is.

Fast forward to 2003. I had been on staff with KCC for 11 years. I started out as the only production person on staff, and now I was leading 12 amazing technical artists on staff. In the beginning, I only knew about audio. During my time at KCC, I shot and edited videos; built and led teams to do image magnification (IMAG) during the service and designed, built and lit the stage on a weekly basis. The technical operating budget had increased 20 times since I started. We moved from the "cafetorium" with the nasty folding chairs to a 1500-seat auditorium with comfortable theater seats and one regional campus. Even though things weren't perfect, God had truly blessed the production ministry at Kensington, and he had blessed me to be a part of it.

I am not an emotional person, but in the summer of 2003 I experienced an amazing worship service that brought tears to my eyes. There was a full rhythm section along with a combined choir of people from our church and a church in the inner city of Detroit. At one point in the service, I noticed the lighting change, and then a complimentary graphic changed at the same time. In that moment, I realized I could clearly hear everything happening on the stage. Every instrument, every voice. It was unbelievable. (For those of you who work with a choir and a rhythm section, you know how hard it is to have tons of open mics on stage and be able to hear anything distinctly.)

I started to cry. I looked around to see if the moment had made an impact on anyone else and saw I was the only one who seemed to be having this specific experience. And I knew this was what God had made me for. Our ministry transformed that worship service. I had a vision of what a technical arts ministry could do to transform something good into something great. I felt God saying to me, "Todd, this is right where I want you."

As it turned out, God was giving me a confidence in who he had made me to be so he could rock my world. In the next several months, I left Kensington and started a new job as the weekend service technical director at Willow Creek Community Church.

This was not something I saw coming or even wanted. I loved Kensington and would have been content to stay there forever. Obviously, God had other plans. My wife and I began a spiritual journey with God, decided to take the chance and step away from everything we loved to watch as God moved and worked in our lives.

I began to see that all the principles I learned at Kensington applied at Willow Creek as much as they were in my first volunteer job at the record deck. I have had the opportunity to see many different churches from all over the world and see that the challenges are the same everywhere. We might speak different languages, but I've found that technical artists in Germany are the same as technical artists in Nigeria or in the United States.

I have been doing production in the local church since I was 13. I don't know everything there is to know. I have been fortunate to be surrounded by people who know a lot more than I do. God has put me in some pretty interesting situations that I have learned from, but many of those have been very tough. As technical artists we have all wrestled with these situations and come away wondering if it is all worth it. Is it possible that God intended production in the local church to be difficult, and there are only a few people suited for the work? And could it be true that I need to figure out how to thrive at it?

I believe God has a plan for you and your production ministry that is bigger and better than most of us are experiencing.

The answer is yes.

I believe we *can* survive it. I believe it *is* worth it. I believe God has a plan for you and your production ministry that is bigger and better than most of us are experiencing.

The following chapters will be connected to my story. They will fill in the gaps of time between when I started in full-time ministry in 1992 to the present. We will journey through my major learnings together.

When I look at the state of production in the local church, I am inspired. There are many churches reaching people, changing lives and families through the use of the technical arts. I am also saddened to see so many individuals barely hanging on, being overworked, their spirits crushed under the pressures of the local church. I hope this book will challenge and inspire you to the possibilities for the technical arts in your church and how you can play a part.

Chapter 1 Discussion Questions:

1. How did you first begin your journey in production?

2. What are some of the defining moments that have shaped how you view the technical arts ministry in your church?

3. What is the biggest challenge for you and for the technical arts ministry at your church right now?

4. What does your production team do really well? Celebrate!

2

ARTIST REDEFINED

art·ist ['är-tist] (plural art·ists) *noun*

Definition:

1. creator of art: somebody who creates art, especially paintings or sculptures
2. performer: a member of the performing arts; a well-known recording artist
3. skilled person: somebody who does something skillfully and creatively; an artist with a basketball
4. cunning person: somebody who is very good at a particular thing, especially something cunning (slang); a rip-off artist

When I was first involved in the production ministry at my local church, the definition of an artist applied to other people, not me. The artists were performers in dance, music, drama, singing, or some other form of visual art. I was just the person who turned knobs, aimed the lights, ran ProPresenter for lyrics—all to support the art that other people performed. It never dawned on me to consider myself an artist.

This perspective resulted in all kinds of dysfunctions in the way I interacted with people on the stage, the way I led my volunteer teams and the way I determined what was or wasn't a valuable use of my time. *I was there to serve.* My needs were not valid. Suck it up and push forward. The real artists are on stage waiting for me to deliver. Don't question, just respond. Get it done.

Statements like these led me to a very passive-aggressive place. Slowly, I became more and more bitter to the point where I was unable to serve anybody. I was not functioning in the way God intended the body of Christ to work, and I was affecting everyone around me—artist or otherwise.

Unfortunately, this is a common occurrence in the production ministry of most local churches.

After years of feeling totally defeated and overworked, God began to change how I viewed myself and the role I filled.

This shift in my thinking became key to growing, not only a thriving production ministry, but to becoming the person God designed me to be.

Defining Mission

In 1999, it seemed like every conference I went to and every book I read said that you could not be effective without having a mission statement and values to support it. I decided it was time to create those for our team of five. Joel was doing audio, Gary was handling the technical needs of our children's program, and Curt did all the video work—pre-, post- and everything in between. Barb had just started as our team's administrator, and I was doing lighting and set design every week, as well as leading the other members of this team.

We met weekly in the new green room behind our newly built auditorium. It was our "secret" location—a place where we could work without interruption. One week I added a discussion about our statement and values onto our agenda. I wasn't prepared for what happened next.

I read through the single page of information containing about 10 values and the mission statement: "To enhance the arts and the spoken word at KCC through the use of the production arts, so that the character of God may be expressed without distraction and life change facilitated for those in whom God is moving." When I was done reading, there was silence. Crickets.

I know what it is like to be in a meeting when somebody puts an idea out there. Just listen, don't object, this person has already made up their mind, so my ideas won't matter. In the silence, I finally let everyone know that this was something for us to talk about, not just accept blindly.

Joel said something like: "I totally disagree with the basic premise of the mission statement. It says we serve other people who are artists. Production done well is an art form, and we need a statement that attracts artists, not people who serve artists."

That had never even crossed my mind. I was going for the distraction-free-let-us-go-unnoticed perspective. Joel and I started getting into it. I don't generally get overheated when it comes to conversation, but he and I were pretty much yelling at each other. Everyone else around the table had to endure the two of us trying to make sense of what we were saying. Unfortunately, we were unable to see eye to eye in that meeting.

Every week after that, some part of our meeting was spent "discussing" our mission statement. These were some of the worst meetings I had ever led. It seemed totally out of control, and I felt like if we couldn't agree on this, how could we move forward? We were working on the foundational statement for our ministry, and we couldn't even agree on what that should be.

I know it doesn't read like the greatest mission statement of all time, but after a month of knock-down, drag-out meetings about what each of us really believed, everyone around our table actually believed it. And we were ready to share who we were and what we were about with our volunteers.

After several meetings like this, it finally dawned on me that some of the statements I had as our mission, while definitely important, were more like values to guide us as we go after the mission. The mission needed to be bigger than providing a distraction-free environment. When it was all said and done, we came up with: "Using the technical arts to reflect Christ to artists, to each other, and to the local church." From there, we let the values help to define what the statement actually meant.

I know it doesn't read like the greatest mission statement of all time, but after a month of knock-down, drag-out meetings about what each of us really believed, everyone around our table actually believed it. And we were ready to share who we were and what we were about with our volunteers.

Those meetings transformed the way I thought and felt about being a technical artist. We weren't just facilitating someone else's ministry, we were *doing* ministry. It was a completely new revelation that affected how I led my volunteers, my staff and how I interacted with other departments in my church.

It became the defining moment for me, not only as the leader in the production ministry, but as a technical artist. This created the foundation for how I would choose to live every following day.

Chapter 2 Discussion Questions:

1. How do you think about what you and your team do as technical people?

2. Do you resonate with serving artists or artists serving together?

3. If you are a leader, are you open to other people's opinions and challenges?

3

PRODUCTION DONE WELL IS AN ART FORM

"Production done well is an art form" means that those who practice the technical arts are artists. Not glorified janitors. Not a group that's subservient to another. Not second-class citizens. Not techies just waiting to be told what to do, which knobs to turn, which lights to aim, which graphics to put up. Artists.

This means I am an artist. I bring something unique to the table that no one else can bring. What I am doing is an art form and, when brought together with other forms of art, is an amazing combination. Does "living out my art" mean that I turn a few knobs? Yes. Does being a technical artist mean I serve the needs of other artists? Yes. Does it also mean that I am still most likely to be the first one in the building and the last out? Yes.

After our meetings around defining our mission statement, nothing physically changed. What did change was that I thought of myself differently, and I communicated what we were about in new terms. Not only to our own teams of production volunteers, but to my boss, to other ministries, and to the church leadership.

I had spent the first 10 years of my production life thinking of myself as a non-artist—facilitating what someone else wanted. Many of us have the gift of helping or serving. It is natural for us to bend over backwards to meet someone's needs. That can lead to saying yes to ideas even when we know they might tank our set-up process. Or thinking of a way to change an element to make it more doable technically, but not speaking up because "I'm just here to serve."

The trouble is that when we don't speak up for our needs, we usually end up doing a less-than-adequate job executing a service. As artists, we need to be able to prepare in order to bring our full selves to the table. I began to notice that when production was done well, when people saw the difference between unprepared production and production as a part of the creative process, we had a voice. We brought the best of production to the table.

This was very different from just serving the artists performing on stage. We were able to enhance what they were doing by bringing the best of our art form to the process. Production, in the setting of a church service or a large venue concert, is an art form with the potential to add immeasurably to other art forms present. The combined forces of music, drama, visuals and production done well can be unstoppable.

At this point, many of you may be saying, "That's right on. Without production, my church couldn't survive. They can't function without us. Finally, someone who understands me!"

Wait. There's more.

Back in the early '90s, we planned a mid-week service to celebrate all the volunteer teams around the church. We divided everyone into their serving teams to figure out ways to communicate to the rest of the church what the teams' roles were. I was the audio/lighting/video director for the service, so all the production volunteers gathered around the soundboard to create a plan. One of the volunteers came up with a great way to describe what we did, and when our turn came we sprang into action. We were meeting in a hotel ballroom and had people stationed at all the dimmer switches around the room. At just the right moment, we turned out all the lights and shouted: "Just try to do this without us!"

The combined forces of music, drama, visuals and production done well can be unstoppable.

It got a huge positive response from the rest of the congregation and became a moment everyone remembered for years. Unfortunately, as my wife says, "All kidding is half-truth." She was right. Our team lived with an underlying bitterness that came from feeling misunderstood. We were working so hard to do what people were asking that we couldn't "serve" any harder. Yet, people still kept asking for more. And more.

So much of our mindset is related to not considering ourselves artists and co-contributors, but rather servants to other artists, subservient to the needs of others. We just do what we are asked, because that is what being a servant means. Unfortunately, this is a recipe for lots of passive aggressive behavior on the part of production people toward most everybody else.

If we want to say we are a ministry of artists, we need to go back to the definition of what an artist is. The phrase that jumped out at me was: "skilled person: somebody who does something skillfully and creatively." This greatly expands the definition of who we consider artists and is the foundation of what it means to be a technical artist.

Let this new perspective lead you towards becoming the technical artist God intended.

Chapter 3 Discussion Questions:

1. What defines your team and how it functions?

2. How has that definition (or lack of definition) affected how your team functions?

4

WHAT IS YOUR ART FORM?

When you expand the definition of artist beyond painter, dancer, singer or musician, it is amazing how many artists there are in the world of production. Thinking of an artist as someone who does something skillfully and creatively helped me to see there are many examples of artists on the production team.

John is one of many artists who come to mind. In some ways, John is the classic tech person. Not only is he really into gear and how it works, but he actually understands how it works and can usually figure out a way to fix it or at least get it fixed. If John has worked on a piece of gear, then the gear will work. It's clean; the solder joints are beautiful; it's clearly labeled. In short, it's a work of art.

John uses his passion for technology and brings all of himself to making gear work properly. So much of what we do would be impossible without the artistry of people like John, who work even further behind the scenes than your normal production ministry jobs.

Take me on the other hand. After a few years on the job, I decided it was cheaper for us to make our own mic cables. I could easily buy the materials separately for much less, but putting a soldering iron in my hand was to be avoided.

After a couple of weeks of consistent use, a solder joint would come undone, and we would throw that cable on a pile to be repaired later. About once a year, we would gather a volunteer team together to clean out our equipment containers and make repairs. It won't surprise anyone I've worked with to learn the majority of cables that needed repair were ones I had made.

It was painfully obvious that I should not be making cables, and we all agreed that I would never make another cable again. In short, I am not an artist in the same way John is.

In the early days at Kensington, we had so little time to do setup that we couldn't afford to have unreliable cables. Not only should I not serve in this capacity, but I was actually hurting our team and the church by doing so.

Fortunately, I don't think anyone's salvation came into question as a result of my mic cables, but for me, serving in an area that God had not gifted me in was painful both for me and for the volunteers around me. When you have a limited time to do setup in a rented space, having to troubleshoot something can eat up lots of time you don't have. This can cause tensions to run high and create possible relational damage. Let me expand with another example using John—the quintessential tech person.

John is great at fixing things, loves to figure out how things work, is quiet and always willing to jump in and help. I am sure we all have volunteers like John. Unfortunately, when we are in need of someone to run audio at Front of House (FOH), we've sometimes had to throw someone like John on the soundboard. We make the assumption that since John knows how the sound board works, he can mix audio. John is an artist, but not when it comes to mixing audio.

Since most churches have a limited number of roles to fill, FOH audio is where these types of volunteers are placed, regardless of their gifts. This mismatch happens all the time in the church. Not only in production, but in the children's ministry with someone who doesn't like kids, or on stage with a vocalist who is tone deaf.

The church is full of people serving in areas where they shouldn't just because it is easy to slot them into obvious volunteer opportunities that need to be filled, while the area where they should serve hasn't been considered. Other times, people get significance from serving in a certain position. They're the ones who think, "My time is too valuable, so I can't be assigned to anything except camera 1 or 2." Then there is plain old church-style guilt. "We really need help and can't do this without you!"

This situation does a couple of things that have detrimental effects on your church. Putting John at front of house will slowly drain him. The pressures of sitting in that hottest of seats will begin to drag him down. Because he doesn't necessarily have the best ear, lots of people will be upset about the mix. (And I'm not talking about those people who are always upset with the mix.)

We need to be diligent about putting people into their sweet spot. For John, it was keeping our gear working. John now spends his time on the things that energize him as an artist. He is filled up in the process, and our church and ministry gain immeasurably from his contribution.

What is your form of art? Are you serving in the right area? While many of us are tempted to find our significance in our performance or position, just because you are into audio, it doesn't mean that FOH is the best place for you.

Thinking of yourself as an artist is key to how to live your life as a technical artist. Equally important is figuring out what your art form truly is. I learned I was not an artist when it came to making mic cables. It has taken me most of my life to figure out what my art form is. You should never stop trying to discern how God has wired you. And when you figure it out, you'll know it.

Before I ever realized there was such a thing as technology, I was a musician. That led to becoming an audio engineer and leading the audio volunteers through setup. I then moved into directing live video as well as shooting and editing videos. After years of that, I became a scenic and lighting designer for our weekly services. And somewhere in there, I drove semi-trucks full of gear.

One of the frustrations was feeling like I didn't belong anywhere. I became mediocre at almost every technical and musical discipline. I knew just enough about all kinds of things to be dangerous, but not enough to be considered great at any one thing. After years of struggling and wondering what God had for me, I have finally figured it out … for now.

My art form is understanding many people's worlds and bringing them together and being able to communicate to technical people and to people on stage. I am an artist as a technical director, a leader of technical people, a bridge between the technical and the artistic.

God has uniquely created you to be an artist of something. It may not be obvious or easy to define at first. Chances are it will be unlike anyone else.

(Hint: it will be). It may surprise you. Are you willing to take the risk and open yourself up to God's unique plan in your life?

Chapter 4 Discussion Questions:

1. What is your art form? What are you specifically gifted to do?

2. Are you serving in the areas you've been gifted in, or are you stuck in a role that doesn't fit you?

3. Think of the people on your team. What is their art form?

4. What are you doing currently that gives you life and fills you with energy? What drains you?

5

BE OPEN-HANDED WITH YOUR ART

Now that you are an artist, it is time to behave like an artist. This means putting your art out there for others to benefit from and even for them to critique.

When I moved to Chicago, my daughter was about five. She loved to paint, and she wanted me to join her. So, I would sit down with her and get out the Crayola watercolor paints, the nylon brush and some printer paper. Pretty soon it became a daily activity for us. We would sit at the dining room table, and we would paint.

As time went on, I got hooked. I started buying real brushes and real paper, and then real paint. (I had no idea watercolor paint came in tubes.) Like every other art form I had tried, I started getting better but realized I would never be great. Then my wife started showing people what I was painting. It was painful to put myself out there for people to express opinions about my creations.

For more traditional types of artists, being critiqued is a normal part of the process. As long as they have been doing their art, they have had to learn the discipline of being open-handed about it. If you are a vocalist, you put your whole self into something and open yourself up to people judging your singing ability. If you are an actor, the same thing applies. You have to just go for it and lay yourself bare. I'm not necessarily saying all "traditional" artists are good at receiving critiques, only that it is part of what comes with public forms of artistry.

Most tech people have not experienced this before in the traditional sense. Every audio engineer has been critiqued on how the audio mix sounds, but we might never have thought of it in terms of putting our art out there for people to critique. As we all know, it can be very difficult to receive criticism.

I work with a great audio engineer, an artist really. His least favorite thing is the older person who comes into the production booth and complains about it being too loud. He takes it as a personal assault. He is reacting to "putting himself out there" with his art and being rejected.

In order for us to get better at what we do, we need to be open to critique. For my friend, his challenge is to know who to take that criticism from. The casual attender who doesn't like it so loud is not someone to get rattled over. If the music director has some suggestions about how the mix should sound, then that is something to work on. Music directors aren't assaulting the person; they have been living with the music for weeks and have an idea of how it should sound.

When you get feedback from someone who matters, what is your first response? Are you defensive? Do you question your worth?

One of the most difficult realities in church work is that we are called to work together. To get the best results for our churches, we need to open ourselves up to other people's suggestions, opinions and criticism.

We are also required to take a hard look at why we are serving in the first place. Is it for the positive recognition we get? Or is it because we are using our gifts in a way that helps build the church? Do we define ourselves by how well we think we perform on a particular weekend? Or by the number of negative comments we get?

Having people criticize our art form can be hurtful, even if the critique is delivered in love. Our art is a part of who we are. However, using our gifts in the context of the local church means we offer our talents to be used by our church—for the common good of everyone.

When you get feedback from someone who matters, what is your first response? Are you defensive? Do you question your worth?

Or do you leverage the critique to help make you better? There is some amount of truth in each critical comment. Are you able to put your ego aside to find that nugget of truth? In 1 Corinthians 12:7, Paul talks about how all the gifts represented in the body of Christ are given by the Holy Spirit for the benefit of everyone:

> *Now to each one the manifestation of the Spirit is given for the common good.*

If you are an artist with lighting, it can be painful to have people say they don't like your lighting design. But you must continue to create for the sake of the church. Will you use the feedback to make you better, or will you let the comments damage your ego?

If you are finding too much of your self-worth from how well you perform, and you also receive quite a bit of negative critique, you might want to look at why you're serving in a particular role. Is it really for the common good, or just your own good? Are you serving out of your true art form, or are you forcing something for your own benefit?

Are you willing to set aside your own self-interest, look at why you're serving in a particular area, and then, if it is for the good of the church, are you willing to let go of it?

Maybe you're serving in an area that really needs warm bodies to make it all happen. If that's you, are you serving with an open hand, or are you letting the fact you'd rather be serving elsewhere sour your experience, thereby souring the experience of those around you?

If you are the only person available to do the lighting design, often times you get more frustrated whenever someone comes to you with an opinion. After a while, you are ready to hang it up. Relax when the critique comes your way. Know that you are doing the best with what you have and that you would love to do better. Tell this to someone in leadership. When the people you work with realize you know lighting design isn't your gift, but you are trying to help out until they find the right person, they will give you tons of grace and support.

Unfortunately, what tends to happen here is that people get tired of giving you the same critique and just give up. Now you're not getting any feedback on your work, which means you become an island and just do what you think is best. You are working in isolation, which is not what God intended. If you're feeling isolated, without much interaction with people about how you're doing, good or bad, you may want to consider why that is.

Seek out someone whom you trust and have an honest conversation. "Am I any good at this?" "Do you have any ideas on how I could use my gifts better?" You will be opening yourself up to the truth, which might be painful, but you will only be better for having asked the question. You will now be potentially freed up to find the spot God made you for.

If you are leading technical artists and you have someone serving in an area in which they think they are gifted but clearly aren't, now is when your leadership gift comes in. Just like you hope someone will talk to that man who keeps singing solos at church but clearly can't carry a tune, someone is hoping you will do the same thing with the sound engineer on your team who needs to be moved out of the mixing seat and utilized somewhere else.

As members of the production ministry, we need to help each other determine where we should and shouldn't serve. Fortunately for us, God has made a place for everyone. 1 Corinthians 12:11 says:

> *All these [gifts] are the work of one and the same Spirit, and he gives them to each one, just as he determines.*

If we have someone serving in an area where they shouldn't, we are doing a disservice to God's design for the church and for that person. The church isn't benefiting from the gifts they should be contributing, and that person is not serving out of an overflow of who Christ is in them.

Many people struggle with finding their identity in where they serve. It will be difficult to move out of the role that feels like the only significant place to volunteer. Fortunately, Paul continues talking about our gifts in 1 Corinthians 12:14-20:

> *Now the body is not made up of one part but of many. If the foot should say, "Because I am not a hand, I do not belong to the body," it would not for that reason cease to be part of the body. And if the ear should say, "Because I am not an eye, I do not belong to the body," it would not for that reason cease to be part of the body. If the whole body were an eye, where would the sense of hearing be? If the whole body were an ear, where would the sense of smell be? But in fact God has arranged the parts in the body, every one of them, just as he wanted them to be. If they were all one part, where would the body be? As it is, there are many parts, but one body.*

God knew we would be wrestling with this issue, so He addressed it. The body of Christ requires that all use their gifts the way He designed them to work. It may not be easy for you to come to grips with the fact that you shouldn't be serving as the main audio engineer for your church, or as the lighting designer or as the graphics person. The journey you will go on from here will make you better, which makes your church better.

If you are in a place of leadership over someone in this situation, God has called you to help this person work through this issue. I am a huge fan of 1 Corinthians 12, but the chapter has less meaning without 1 Corinthians 13:1.

> *If I speak in the tongues of men and of angels, but have not love, I am only a resounding gong or a clanging cymbal.*

Basically, God calls us to use our gifts in love. If you are a leader of technical artists, you need to remember to use your leadership gift with love, which includes helping your volunteers find the best area for them to serve. People on your team can choose not to go through this in a healthy way, but God calls you to move this person out of the position that they were not designed to fill. Let God handle the other person's response.

Were you created to be the best graphic designer at your church? Were you meant for video directing? Does God receive pleasure from the offering of your gift to Him? Are you open-handed with the artistry God has gifted you with? Are you serving in that area?

Realizing you are an artist is foundational to how you function within the body of Christ, the local church. Realizing what your art form truly is and how you use it is fundamental to being the technical artist God created you to be.

Chapter 5 Discussion Questions:

1. Are you serving in the area you are gifted in? If not, in what places could you contribute more accurately to whom God made you to be?

2. Are you open to critique or would people say you are defensive about your art form? What are some ways you could grow in this area?

3. Are there members of your team who would serve better in different roles? What would be a loving way to communicate this? What would the benefits be for these people? For the team?

6

START WITH THE FOUNDATION, THEN BUILD ON IT

So now you're an artist serving alongside other artists, and everything is amazing! If only it were that simple.

The next sections of this book will dive into the practical side of what it means to be a technical artist in the local church. But first we need to set up a framework for what it looks like to be a technical artist and then determine how we use that art to collaborate with others.

Many years have gone by since those knock-down, drag-out meetings about our mission statement at Kensington. Since then, as I have tweaked it in my mind, and worked with different teams, it has morphed into something slightly different.

"To create life-changing moments through the fusion of the technical and creative arts."

This new mission statement makes one giant assumption and captures a few key ideas that are critical to the conversation moving forward.

The giant assumption is that God changes people's lives. And he does. But by themselves, production or the creative arts, or the level of fusion between the two cannot change people's lives. For whatever reason, God has chosen to work through his people to bring heaven to earth, and so in the spirit of this idea we want to join God in creating these life-changing moments.

Whew! Now that's out of the way!

Next to that assumption, I believe that God created us to live in community with each other, and for the kingdom of God to be realized on earth, technical artists and creative artists need to learn to work together. Production by itself is pointless. And in today's reality, most creative ideas can't effectively be accomplished without the use of the technical arts.

Many of the life changing moments we try to create in our local churches require both groups of people to work together. Collaboration is essential. And from my perspective, this is where many things break down in the local church. These two groups of people couldn't be more different, yet we are called by God to create these moments together.

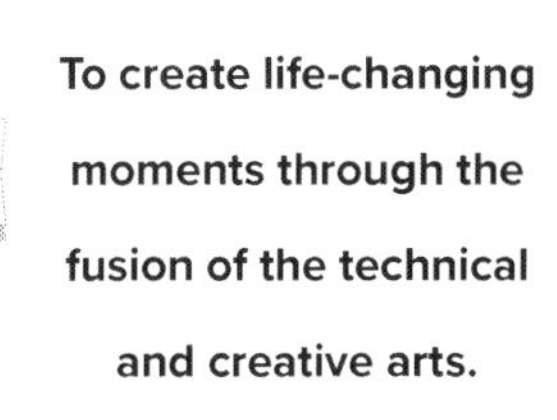

Earlier in my journey, I would have said this was the most important issue we need to talk about. It is the elephant in the room no one wants to acknowledge. While I still believe this, there is an earlier step in the process that needs to be addressed before the fusion of creativity and technology can succeed.

The Foundation

In order for us to have crazy amounts of collaboration and "fusion," as the mission statement says, we must first build a foundation. This foundation is made up of the basics of production. To use a football analogy from Vince Lombardi, we need to start with "this is a football" basics. Without the fundamentals of production, without nailing the details associated with technical excellence, without understanding signal flow, it doesn't matter how much you want to fuse your art with someone else's, it won't happen. If I'm on the side of the creative arts, why would I want to collaborate with someone who can't stop feedback from happening every time the pastor gets up to talk?

Since I am a former engineering student, I dig graphs, so let's take a look at one that describes what I mean:

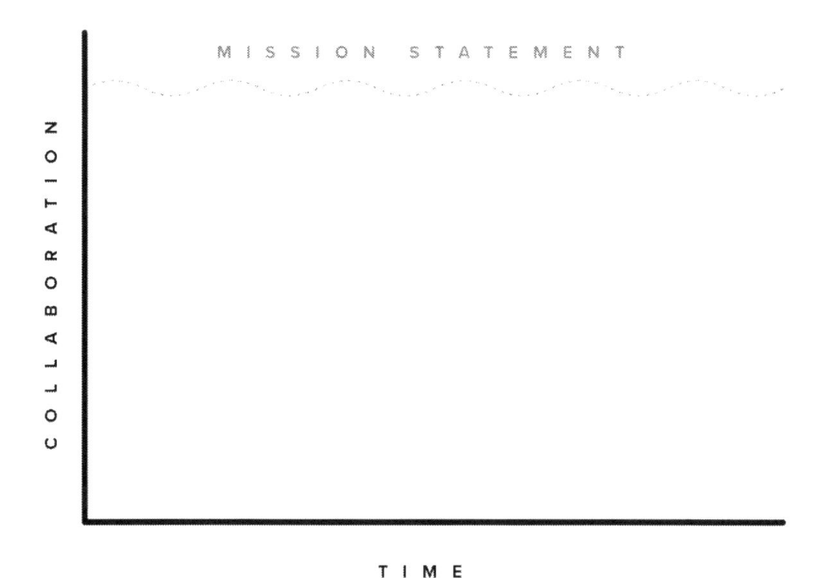

The x-axis represents time, and the y-axis represents the amount of fusion and collaboration between the technical and creative arts.

If we look at the mission statement one more tme, it hovers very close to the top of this graph ... somewhere that seems quite unattainable. A lofty goal and not possible. At least not for me and my situation.

Like any good mission statement, it represents something that is aspirational. It is something we are striving for, it is the direction we are attempting to head in. So, if it seems so far out of reach, what are the steps to start reaching this ideal?

In my experience, the way to reach these heights is achieved in two parts. The first part is necessary before we can begin on the second. One is universal and the other is situational. Let's break it down.

The first step to reaching this goal requires a foundation that is made up of many tiny, yet necessary parts. These things are true regardless of the creative art that we are trying to support. They apply to an event of 10 people or 10,000 people. They are as true for a talking head with a PowerPoint presentation as they are for the Super Bowl half-time show. They apply to a mega church as much as they do to a church of 100.

The production foundation is made up of small pieces: mics being on at the right time, lights pointed at the right things, or readable graphics for the songs. The foundation is made up of many small things that, when combined, don't hinder the content that's coming from the stage. Production elements done well appear invisible.

For us to attain true collaboration, we need to nail the foundational elements of production that only production people know and care about. Unfortunately, we live in a world where even the smallest mistakes can be amplified, to the point where everyone knows that we've blown it. For most in the audience, they know that someone isn't paying attention, yet they have no idea what to do about it. That's where you come in. You are the artist of live production! You know what needs to happen to make it amazing.

This foundation isn't automatic, and it requires lots of tenacity to keep the level up. It isn't romantic or sexy. It's about wrapping cables properly at the end of a service. It is about arriving early to make sure the videos are checked before the room fills up with people. It is the most unseen part of what we do, but without it our missteps tend to be noticed by everyone.

For us to get to the levels of fusion that we dream about, we have to be nailing the foundational parts of production. Nothing builds trust with other people like doing a job extremely well. Even though most non-technical people don't have a clue what you do exactly, most of these same people know who's doing the basics of production well and who isn't.

Many production people want to skip over these mundane parts of production and go straight to the cool and amazing parts. However, until you've proven that the boring and unseen parts of production are happening on a regular basis, you haven't earned the right to go after cool and amazing. Often times, jumping immediately to cool and amazing probably means you don't fully understand your programs, and you are just piling on amazing technology that has nothing to do with the content ... which seems backwards.

How many of you have met someone who wants to volunteer on your production team, but only wants to mix and not do any set up? Generally, someone like that can't be trusted with the keys to the audio console if they haven't shown an understanding of the basics of audio or even a willingness to do the mundane parts of audio.

Part 2 of this book is all about what contributes to this foundation and how to go after the production side of the equation. It is a set of production values that have proven themselves over the years in every production I have been a part of. These values are broken up into small chapters, so they are easy to digest one at a time.

Now Build On That Foundation

As we begin to get a handle on the foundation of production, the door to collaboration will begin to open. So, you math types are wondering, what's the deal with the rest of this graph? Let's expand it to show how we can increase our level of fusion with the creative arts.

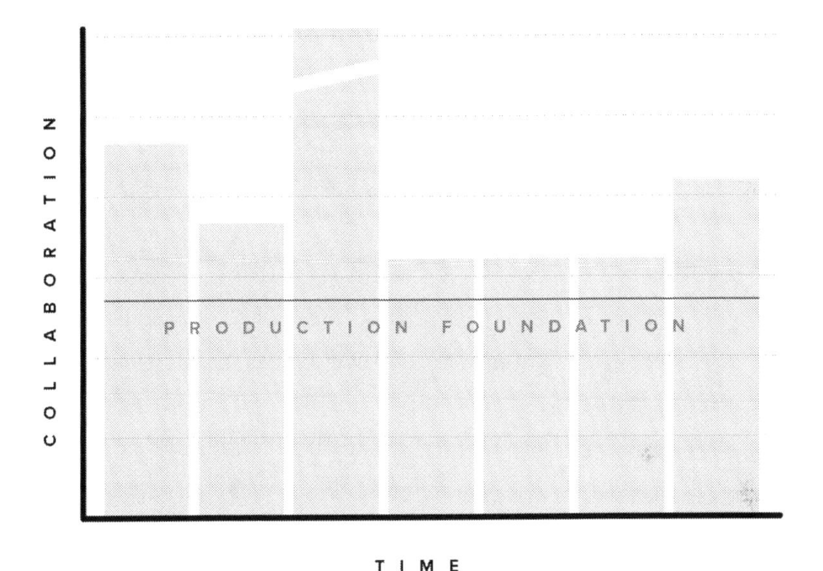

You'll notice that, week by week, the amount of fusion required varies. This is because on some weeks, the content calls for a lot of technology to make it happen. On other weeks, if the mics are on and the lights are working, nothing more is necessary. This is the part of the graph that is completely dependent on the requirements of each event. It is also entirely dependent on you to nail the foundational parts of production. You can't have cool amazing fusion of technology and creativity if you don't get the basics nailed down.

One important thing to realize about this kind of fusion is that it requires trust that has been built up over time in the foundational parts of production.

You can't create together if there isn't a track record that shows you can be trusted to do your job. If you can't put the right graphic on the screen at the right time, why would anyone listen to your idea for using technology to make something better? If nobody is listening to your ideas, collaboration can't happen. And if collaboration isn't happening, you need to reexamine how well you are accomplishing the basics of production.

And if we shouldn't be talking about this part of the graph, you will just be doing technology for the sake of technology. You'll be doing it because the church down the street is doing it. You'll be using the shiny, new technology just because it is shiny and new. Unfortunately, many production folks love to push the envelope of production whether or not it is the right thing for the moment. The drums sounding amazing, some cool new font, your favorite hot pink gel or a crazy handheld shot aren't always what the service needs. Sometimes the things we love are done out of context and it won't work.

You can't have cool amazing fusion of technology and creativity if you don't get the basics nailed down.

Without technical and creative fusion, the technical by itself is pointless. No one will come away from a church service with their lives changed by the lighting. As a friend of mine says, people don't leave the service humming the kick drum. However, very little can be done without the use of technology, and we therefore have a unique opportunity to go after this second part of the graph.

In Part 3 of this book, we'll talk about some essentials to building into the trust we've worked so hard to earn from Part 2 and look at how to maximize collaboration for the sake of the whole experience.

Fusing the technical and creative arts is not easy. The journey will require large amounts of tenacity and patience. And if you thought the tenacity and patience required for building the production foundation was intense, brace yourself.

When you're taking care of the production basics, you only have to worry about yourself and your like-minded team. You understand what needs to happen production-wise. But as you start collaborating with people in the creative arts, you enter a realm where people don't generally understand what we do or how it gets done. (Of course, the reverse is also true. We don't totally know what is involved with creating content week after week.)

I don't believe I've ever met a production person who didn't long for the experience of maximum fusion between the creative and technical arts. This is a big reason we got into this in the first place. We want to do something amazing and life-changing.

Chapter 6 Discussion Questions:

1. How well does your church handle the foundation of the technical arts? In what areas could you improve?

2. Are you using technology for technology's sake, or is true collaboration happening between the technical and creative artists?

3. Does your team have a clear mission? If not, what are some characteristics that currently define your team? Could they be the building blocks to a defining statement?

PART 2

PRODUCTION VALUES

7

TENACITY IN THE BASICS

Building a solid foundation for the technical arts at your church requires gobs of tenacity. If you think about the foundation of a building, most of it is underground and will never be seen. But without it, the building couldn't stand. Even though nobody sees what goes on down there, if corners are cut the building will eventually come down. Yet, by building a solid foundation, there is no telling what can be built upon it.

In the world of production, most of what we do goes unseen. You are the first in the venue getting things ready, and you are the last to leave. There are countless hours in the editing suite getting things just right. Sitting behind the lighting console, checking and rechecking the sequence of lighting cues doesn't happen by itself. Testing each mic line and instrument cable has to be done so that we know everything is working before we start rehearsal.

Most of us can relate to how tired you can be at the end of a long run of rehearsals. Do you stay and clean up now, or leave it for later? Do you watch the video one more time to make sure the edits line up with the audio? Do you troubleshoot a problem until you understand what happened and how to make sure it doesn't happen again?

In the world of production, most of what we do goes unseen.

When I started shooting and editing videos, I learned this lesson the hard way. After I finished a project, I would start transferring it to tape (remember tape?) while I got up and stretched my legs. Because of my stretch break, I didn't watch the master until we were in the service. Inevitably, there was a glitch or a piece of bad audio. It didn't take me long to realize that I needed to watch the final transfer ... all the way through. In one instance, it was a 1–1/2 hour final edit and it was 3 a.m. Do I watch the whole thing, or do I take a nap? If I want to make sure it is done right, I need to watch the whole thing. (By the way, I watched the whole thing!)

These are all examples of tenacity in the basics. Since there is no one around to see that yours is the last car in the parking lot, what you are doing is definitely unseen. Foundational.

Without a commitment to building a great foundation, everything else is worthless. What you are able to accomplish through your production ministry will be built upon the time and tenacity you spend on the basics of production.

Every now and then, especially after a ministry fair or something similar, I am in contact with people interested in serving on the production team. Once I make the initial contact and weed out the people whose mom signed them up, or the people who thought the production ministry had something to do with manufacturing, many of the people left are interested in only mixing FOH or only running the jib camera or only being the lighting designer ... the cool jobs.

However, much of what production is all about happens way before you are sitting behind one of these positions. Typically, I tell people that there's a chance they may never mix in the main auditorium, but we have lots of other areas where they can serve. Stage set up, tear down, aiming lights, setting up risers, etc. This is when you can normally find how serious someone is about serving in production.

The basics matter. Small details are important. Amazing is the sum of the mundane.

The foundational areas are where you see someone's heart. It is where you see how far they are willing to go to create an environment for people to engage with God. This is where you see whether or not someone is there to serve others or is just serving their own needs.

Unfortunately, many of us derive our worth from what we do, not who we are in Christ. This can get in the way of someone's willingness to serve wherever they are needed. You need to remind yourself constantly that your worth doesn't come from wrapping cables at 11 p.m. after a huge mid-week service. Talk about tenacity. If you aren't going back to the source of your worth, working on the basics of production can eat you up and spit you out. Is this why so many tech people in the local church quit from burnout?

Are you willing to do your best work on the basics? If you aren't tenaciously going after the small details that make up the foundation of production, you haven't earned the right to work beyond this. If your wireless mics are constantly failing, and you aren't coming up with a way to solve the problem, you shouldn't use wireless mics. If you are having trouble framing shots correctly, you shouldn't be doing IMAG.

Amazing production is built on the shoulders of the basics. Amazing production is a combination of all the tiny details poured into the foundation of the basics. There are no shortcuts. No one is going to cheer you on for your great foundational work. Going after excellence in the area of the basics requires tenacity. Without it, the glamorous parts of production will eventually crumble.

The basics matter. Small details are important. Amazing is the sum of the mundane.

Chapter 7 Discussion Questions

1. What are some shortcuts you are taking now that will eventually catch up with you?

2. Are there any weak points in your foundation? What are they?

3. What are some ways you can strengthen your production foundation?

8

BABY STEPS

When most people visit Willow Creek Community Church, they are amazed by the sheer magnitude of the place. When viewed from the standpoint of technology, it can be easy to be blinded by the scale of all that is going on. In reality, Willow Creek hasn't always had that much gear. If you were to look at some pictures from the 70s, you'd see some pretty scary setups.

Where they are right now is a function of taking small steps, continuing to add functionality and possibilities along the way.

As someone who loves to try new things and loves new gear, I know it can be tempting to want to jump over all the steps that Willow Creek took and just make the leap from where I am right now to the top of the pile. In reality, this is a plan waiting for disaster.

So, how do we aspire to something more for our churches without getting discouraged by the huge gap that exists between where we are right now and where we'd like to be? Take baby steps.

In my opinion, "What About Bob?" is one of the classic Bill Murray movies of all time. Bob is a super dependent patient of Dr. Marvin who ends up driving the doctor insane when he visits him on his family vacation. One of the main ideas from the movie is Baby Steps—the concept of facing your fears one small step at a time. Essentially, any fear Bob was facing, like being afraid of leaving his apartment, he could overcome by taking small steps toward the door, one at a time.

In a similar way, any big change we need to make starts with small steps toward a solution. Steps we can handle and can experience success with.

It Better Work!

As tech people, much of what we do requires equipment, and that equipment requires money, sometimes lots of it. When we are asking our church's leadership to spend money on gear, they need some level of assurance that it will work. If we are set on taking giant leaps towards more and bigger, can we be sure that it will succeed? It better.

If you ask for lots of money for more equipment and it turns out to be more than you can handle, the next time you ask for money, the answer will probably be no.

Many years ago, we had some volunteers who were into satellite technology. They owned a bunch of KU band gear, and the living room of their house was full of racks of stuff to watch whatever KU had to offer.

One year, Kensington was a host site for the Global Leadership Summit, a leadership event broadcast live around North America from Willow Creek. We had volunteers who thought we'd get a better-looking signal if we skipped the small dish that came with the Summit and used a KU band dish. They had the necessary equipment and wanted to donate it, but we needed to spend about $500 to make the dish work with our system.

For those of you in production, you know that $500 doesn't go very far when you are buying gear. From my standpoint, it didn't seem like a ton of money for a great result.

Well, if you didn't know, a KU dish is pretty huge and difficult to miss. Especially when it is sitting out on the grass in front of the church. While most of the $500 gear we buy fits into one rack space, this was $500 that could not be missed by even the most technically ignorant.

Soon after we set this up, the senior pastor called me into his office to ask me what we were spending all the church's money on. He was less than happy. I explained to him what the plan was and that we'd test the difference in the video signals and let him see the difference.

When the day of the test came, I admit that I was praying like crazy for us to be able to see a difference in the two video signals. I imagined the trust I was going to lose if this didn't work. Fortunately, the difference was obvious, even to the senior pastor. Crisis averted!

After the dust settled, I started to reflect on what would have happened if the test had failed. I could imagine my budgets getting cut, and the ability to continue to grow production to keep up with the growth of the church would be severely hampered.

It made me very aware that there are only so many times you can take steps toward doing more and not having it succeed, before trust gets eroded. That $500 satellite test brought to the surface the reality that I needed to be sure I was only asking for things we could do well.

In reality, the history of production at the churches that use production well is a story of taking small steps that lead to small successes. I have no doubt that buried in the details of the last decades at these churches, there are a few production missteps, but the general trajectory moved them in the right direction.

Since we all have limited resources, it is important to spend them as wisely as possible.

When I was a few years into the lighting part of my career, moving lights started to become a big deal. There were quite a few people on the team, my boss included, who wanted us to look into buying moving lights.

I pushed back pretty hard because I knew that I didn't have the time necessary to devote to programming them. (Did I mention I was also shooting and editing all the videos at the time?)

I finally gave in, and we bought two Martin Mac 250s(!). Except for Christmas and Easter, I never used them. They collected dust. And while they collected dust, I kept thinking that we hadn't been ready for this step. It may have been time to move in this direction, but we didn't have the capacity to succeed at it.

Looking back, I should have continued to push back, and I should have figured out a way to spend the money on something else—something that would have moved us to the next step we were ready to take.

Since we all have limited resources, it is important to spend them as wisely as possible.

Sometimes saying no to something really cool is the right answer.

There are two sides to the idea of baby steps. One is not to get discouraged when you see a concert or a church that is way ahead of where you are from a technology standpoint. They didn't get there overnight.

The other side is not to take a step beyond what you can handle. While there is something good and essential in stretching yourself toward something new, by overreaching you are opening yourself up to failure and for a dip in trust with your leadership.

If you say, "We really need this," and then don't deliver, you've begun to erode the trust that you've spent so much energy to build. And it's more difficult to regain trust than it is to build it in the first place.

Chapter 8 Discussion Questions:

1. Do you have any examples of asking for more than you can handle? What could you have done differently?

2. What is your next baby step? Can you make a case for success?

3. Are your next steps characterized by helping the church grow or just the production department?

9

A CLEAN STAGE IS A HAPPY STAGE

When I was first married, my wife would move my stuff. I'd have little piles of "junk" (in her words), located in each room of the house. In my mind, they were there for a reason, and I knew where everything was. She didn't see things in quite the same way. Those piles drove her crazy. She couldn't understand why I didn't have a specific place for each thing. My version of organization didn't seem very organized to her.

However, when I tried to pick up her piles of stuff, she had a different perspective. My own unique style didn't make sense to my wife, and hers didn't make a ton of sense to me. We discovered that in the public spaces of our house, we needed to have a system to keep things ordered and looking good. From there, we each had a corner of our bedroom where we could put whatever we wanted. If she wanted to pile up her clean clothes for weeks, that was her choice. If I wanted to hang onto those old VHS tapes of behind the scenes footage of some production I did in the early 90s, that was my call. If I had trouble parting with old pairs of glasses (because you never know when you might need a backup), I could knock myself out. (These are hypothetical examples, of course.)

For each of us, our "organization" was invisible to the other. It made sense, and we never had to think more about it. To the other person, it was an eyesore.

Like your living room, your stage is a public space. Not only are there volunteers on your team who use that space, but musicians and pastors spend a good deal of their time on that stage. All the people in the congregation are focusing their attention on the stage. This is your common area. This is the place everyone sees.

Most of us are in a hurry to get things set up and working. I can remember thinking that I didn't have time to make things as neat as they should be. I've got to get this thing up and running! Who's got time to sweep the stage or keep the cables neat or pick up after last weekend?

In my mind, there are two sides to this conversation. One involves being organized, the other involves creating an environment.

Get Organized

If everything is helter-skelter, how easy is it going to be to troubleshoot a problem? When we are knee deep in rehearsal and something isn't working, is one of our volunteers going to be able to dig through the rat's nest that we've created? Are things organized in such a way that your team can find what they need in the heat of the moment? And what happens when you want to take a vacation?

Keeping your area neat and clean helps keep things in their places and easy to find. When you let it get messy, the organization makes less and less sense to other people. Going back to the piles of junk I used to have in every room in our apartment—I knew exactly where everything was, but nobody else had a clue.

The minute other people are involved in what you are doing on stage, there needs to be a way for others to understand the order. This can't happen if your stage is a disaster.

Take the time to make it neat and organized. Put those cables away after each use. Develop a system to keep extra light fixtures well ordered. The initial push will take some doing but maintaining the system will be worth the few extra minutes.

Creating an Environment

The other part of keeping the stage clean is about creating an environment for the people on the stage and for the people attending your church. Do you want people to feel like they walked into a house from one of those hoarder reality shows? Not to over spiritualize this, but God is a God of order. Paul talks about this in 1 Corinthians 14:33:

> *For God is not a God of disorder but of peace.*

Order reflects part of who God is. Taking time to be intentional with how cables are run matters. Figuring out a place for everything is important.

We want our teams to feel comfortable and at home when they are on the platform, to feel like there is an environment they can create and serve in without worrying about tripping over something.

It is amazing to me to watch people treat a messy space poorly, while they tend to treat a clean, orderly space with more respect. A messy stage gets messier. A clean stage can still get messy, but if everyone understands that "normal" is clean, it is easier to hold everyone to the standard.

When your team is on the platform or people from your congregation see what the stage area looks like, what impression does the stage make? Is there junk everywhere? I worked for a pastor once who judged a person's character by how the back seat of their car looked. I'm not sure I would go that far, but what does the stage say about you and your team? Does it show people you treat your equipment with care or that you disregard the tools that have been entrusted to you?

Chapter 9 Discussion Questions:

1. Go to your stage area and see who can find the service cue sheet from the longest time ago.

2. What is an area that has become invisible to your team that needs to be fixed?

3. What process can I set up so that the team can stay organized even when I'm away?

10

USE WHAT YOU HAVE

The world of production is always changing. The technology at our disposal is increasing while prices are coming down. With so many things evolving, it is easy to get caught up wanting to have the latest and greatest. You want to replace something you have now with something that can do more, or something that might put your church in the same category as the "big dog" churches.

While this might not be exactly true for you, it tends to be the perception that non-tech people have of us: Our striving for more gear and the newest version of whatever we currently have is about just that. We like new and better because it's fun to have new stuff. As a result, people's perception of tech people is that we are completely disconnected from the idea of ministry, and we're just about the stuff.

How do we change this perception?

Use What You Have

The next audio plug-in or the next piece of gear is not always the answer. I can remember blaming the gear for the mix being bad or thinking if we only had a different CG computer, the graphics wouldn't be late. However, blaming the gear for why things aren't going well is generally not a good place to start.

Every time you blame the gear you are not taking responsibility for the performance of your team or yourself. I'm not saying that the gear isn't at fault some of the time, but it should never be the starting point for why things aren't working. Take responsibility for why things aren't working and then take responsibility to make them better. More often than not, mistakes can be corrected by changing the process: adding more time to set up, simplifying the process, getting a second set of eyes on the graphics.

I can remember when a new piece of gear was installed at one of our campuses, and the first weekend it didn't work. The campus pastor called immediately and wanted to replace the equipment. We asked if they had checked it before the service. They hadn't. Having decent gear is fine, but if you haven't watched the video, or tested the mic before the service, how will you know for sure if it will work? New gear is not the answer.

Take responsibility for why things aren't working and then take responsibility to make them better.

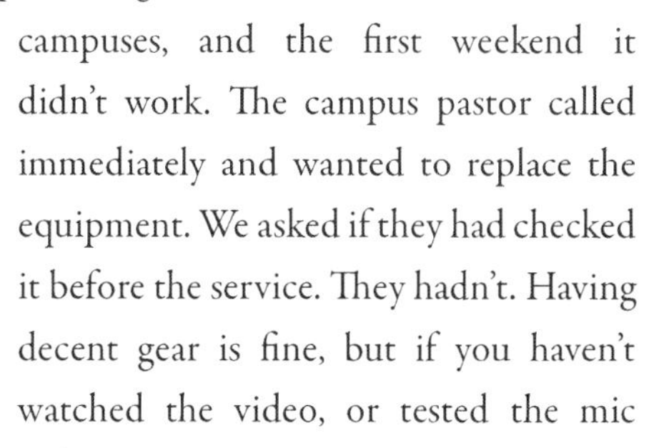

Regardless of the size of an event, or the size of the budget, there is a lot of space for excellent production, even with less-than-perfect gear. I've been a part of some pretty amazing productions, and all we had were homemade speakers and a bunch of SM58 microphones. It can be done.

Back in the day, I used to attend conferences where the production level was amazing. I would be inspired for a couple minutes then get depressed because I could never hope to achieve similar levels at my church. I was guilty of gear envy.

After a few years of dealing with this let down, I started thinking. Instead of just giving up, how could I take what I already had and use it more and better?

Get creative with what you have in front of you. While having new and better equipment can be a good thing, it can also come with more problems. Learn how to be excellent with what is in front of you now.

Get the Most Out of Every Piece of Equipment You Have

Get out the gaff tape! Wring every last ounce of usefulness from the stuff you do have. How many times can you repair a light fixture to keep it functioning?

How can you patch audio this week to achieve the desired result? Will putting another drywall screw in the stage deck keep it functional for a little while longer?

However, if a piece of gear actually isn't working, then it is probably time to replace it. At this point, it's time to let your leaders know the risks involved and get them to help you make a decision about what to do.

The main PA at Willow Creek is pretty complicated. It is a big room, with lots of speakers in different locations that all need some form of processing. One weekend, we heard a pop and then part of the system went down. After troubleshooting and figuring out where the problem was, we fixed it. However, along the way we discovered that they didn't make parts for this gear anymore, and there were none that we could rent in an emergency.

At that point, I went to my boss and explained the situation. We could make it work, but I felt like we were on borrowed time. Since making sure people can hear the gospel message is what our church hinges on, I felt like we needed to look seriously at replacing this.

We had a plan to keep the old system working, and we had a plan to replace it. I involved the decision-makers at my church to help assess the risk and make a decision. Being ready to squeeze more out of the system speaks volumes (pun intended) to the leaders at your church. Being ready to do your best, even with less than ideal gear, says that you are a team player and willing to do whatever is necessary to advance the ministry of your church.

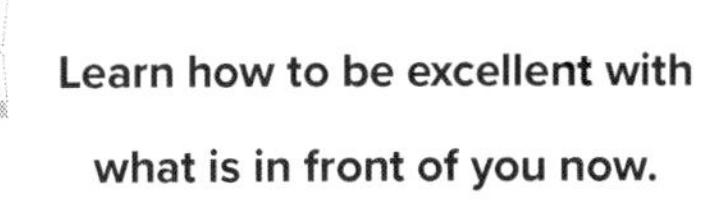

Continual duct taping of gear without any communication to your leadership is a mistake. I recently worked with a church that was great at keeping old gear working. However, no one in leadership knew how bad it was. When wholesale system failure started, the fact that there was so much tape holding stuff together didn't seem like such a great value.

Share the responsibility and the risk with your leadership. Don't carry it yourself. Keeping others informed about the condition of the equipment along the way is key to building trust. Decide together, not alone in a vacuum.

Don't Wait to Stretch Yourself

In a similar vein to getting the most out of every piece of equipment, don't wait on new gear to try something new. How can you take what you have and do something new? How can you leverage the tools you DO have at your disposal for kingdom impact? Being ingenious with what you have can be a creative challenge. Think differently about what you have and see what you can come up with.

You may not be doing things by the book, but who cares? Try it anyway. You might learn what doesn't work, but you might also learn that your gear is capable of much more.

Friends of mine who do production at their small church wanted to do a double-wide video image but couldn't afford projectors with that feature. They ended up experimenting with what they had and figured out a way to make it work. Is it perfect? No. But 90% of the time you'd never know that a couple small pieces of aluminum rod were used to make the edge blending work.

Surprise. Money Doesn't Grow on Trees

At least, it doesn't at my church. Until it does, funds will always be limited. There isn't enough money in the world to buy new and better at every turn. Learning to live with what you have is a normal life skill that we all need to master.

Don't Get All Passive Aggressive About It

If you have reached the end of the usefulness of your lighting console, don't just shrug your shoulders and say, "I'm trying to get the most out of our gear, so we can't expand into that area." This is counterproductive.

There is definitely a point where you will reach the end of the usefulness of a piece of equipment. Don't shy away from asking for more gear if it's needed.

When your team and your leaders see you responsibly stewarding what

God has entrusted to you, their respect and trust will grow. When you leverage all you currently have faithfully, you will be entrusted with more.

Jesus talked about this principle in the parable of the shrewd manager in Luke 16:10:

> *Whoever can be trusted with very little can also be trusted with much.*

Chapter 10 Discussion Questions:

1. Think about the times you've blamed a mistake on the wrong equipment. Was that really the case?

2. How can you get creative to keep equipment running?

3. What is something new you can try without buying new gear?

4. How can you include your leaders in the state of your equipment and share the burden of responsibility with them?

11

SET THE TABLE

Remember that time you decided to have the whole extended family over for Easter brunch, even though you were busy trying to pull off an Easter production?

One of the things I love about having people over is that it causes a flurry of activity to clean up, pick up and organize our house. For many of us, the mess in our homes becomes invisible to us, and we let things pile up. Having people over means that they are going to see the mess that we've been living with for who knows how long.

On the other hand, one thing I don't love about having people over is that it takes so much work to get the house ready and presentable. But it must be done. And it must be done before anyone arrives. I don't want people walking in while I'm still cleaning up. I want to be relaxed so I can enjoy our guests.

The guests have a certain expectation that they will be taken care of. They are anticipating an enjoyable, relaxing time.

My friend Marty O'Connor, a fellow technical artist, taught me that the principle of being ready for your guests applies to what we do as technical artists. One of the key factors in being prepared is that we have the table set, so to speak. When our counterparts on the stage arrive, everything is ready for them to dive into the task at hand.

The musicians, vocalists and speakers who have a task to perform on our stage have many things going on in their heads and hearts as they prepare to lead the congregation. Our job as technical artists is to have everything ready for them so that they can concentrate on the part they need to play.

This means that line check has already happened, that the lights are aimed, and the graphics are correct and ready. The goal should be to have everything prepared before the band walks in the door, much like a dinner party. If a guitar player has to go digging around looking for a music stand, she isn't able to focus on what she does best—play guitar.

I'm not suggesting that musicians should be above helping out, but when you boil it all down, it's the production team's job to have everything prepared and waiting for people to walk up and do their thing.

There was a season when we were never finished with our set up on time. We were meeting in a high school auditorium, and we had a lot of stuff to get ready in a very limited amount of time. In order to have the table set, we started a tally of how many times we were ready to go, with our feet up, waiting for the band to arrive. It became a contest to see how many times we could do this.

I'm sad to say that it happened less than I wanted it to, but it gave us a goal to shoot for and an awareness of getting the table set before our "guests" arrived.

As technical artists, our job is to take care of the technical details of our services. Our pastors and worship leaders should be able to walk in and only worry about what they need to take care of—not whether the graphics will be ready.

To take this up to another level, what if you spent some time figuring out how certain people like things to be ready for them? The drummer only likes to use one tom, not three, so our team takes the time to make it so. The senior pastor always likes a small table for water to the right of the podium, so it's there.

For those people who have ever stayed at a Disney resort, you know that the service is outstanding. A friend was telling me that after housekeeping cleaned their room, his son's stuffed animal was moved around every day and posed in a fun way: brushing its teeth, looking out the window, watching TV. Was the room clean? Sure. But the housekeepers took the experience to another level by spending a few extra minutes to show some thoughtfulness.

Chapter 11 Discussion Questions:

1. Are you and your team ready to go when people arrive on stage?

2. What do you need to change to make sure the table is set and ready to go?

3. How can you go out of your way to create an unforgettable experience for your worship team, your pastors or other creatives?

12

DISTRACTION FREE

Have you noticed that in any movie about life in high school, there is always a scene in the school gym or auditorium that involves audio feedback? Someone comes up to the mic, taps on it, says "is this thing on?" and we get a little dose of feedback.

Why is it this way in the movies? Unfortunately, it is the norm. I've actually been to a few events at my kids' schools, and I can say this is pretty accurate. In an "amateur" setting, we expect it. However, when you go to a concert or to a "professional" event, it is a rare occurrence. When we are watching the Grammys on TV, Twitter lights up with critiques about the mix, or a mic not being on ... because we expect more from the Grammys.

So why is amateur-level production acceptable in many churches? Why is audio feedback accepted as part of doing church?

At the base line, production is a transparent way to amplify what is being communicated. When it is more opaque, it is getting in the way of the message. When people notice the production elements, they are distracted from what they are supposed to be paying attention to. In the context of our local church, that is the gospel message of Christ.

For those of us in the local church, that gospel message is the whole reason for production to be involved in the first place. Our equipment and our volunteers are in place to help communicate the gospel message. Compared with other "professional" settings, I would put our mission far above theirs. For the Super Bowl, they are doing great production work so they'll get hired again. We're doing it for a different reason.

Inevitably some of you are saying, "If I had the resources of the Super

Bowl, I could do excellent production." Or "If I got paid, I could devote more time to excellence." I would say you are wrong.

The very first goal of what we do as technical artists is to create a distraction-free environment whether you have all the resources in the world or hardly any. It doesn't matter how much you know about the latest technology or how many new gadgets you have in line between your mixer and the microphone. If people are distracted from the message, it is all for nothing.

There are two ways we can be distracting as technical artists. One that we've already talked about is eliminating potential distractions: audio feedback, wrong graphics up at the wrong time, lights aimed improperly. And second, sometimes we pile too much technology on top of our services, making it more about the audio, video or lighting than the actual content of the services. An audio mix that is too loud for a particular moment. Too much haze in the room. Camera shots that require the camera operator to be in the way of what's happening on stage.

Too much production that doesn't match the intent of an idea can be as distracting as missed cues and a mic that feeds back.

How can I make it so that people focus on the message and don't even notice what the production team is doing?

One year after a Christmas service, I went to a Christmas party where people were talking about the service. When I asked them what they thought of it, the number one answer was that the lighting was impressive. Not that the service was meaningful, or that the message really hit home. If lighting is the only thing someone can say about a service, then we probably overdid it.

The goal of amazing lighting is to create an amazing service, not amazing lighting for its own sake. If people only see the lights, they've been distracted from the primary message of what we should be doing.

Is what we are bringing to the table with technology making our services better or is it distracting people from the message of our services?

What do I need to do to create the least amount of distraction possible? How can I make it so that people focus on the message and don't even notice what the production team is doing? There are many facets of production and many values that we need to hang onto, but the idea of creating a distraction-free environment is at the top of the list. How can production get out of the way so people can be free to engage with God?

Regardless of the resources you have at your disposal, if you aren't eliminating distractions, you probably don't deserve the tools you have been given. If you take someone who has mastered the concept of creating a distraction-free environment, it doesn't matter how big a PA is, they will use the tools at their disposal to make the audio transparent for the event. On the other hand, if you take someone who hasn't mastered the concept of transparency, it doesn't matter how much gear they have, the equipment will not eliminate distractions for them.

Let's change the stereotype of feedback and people tapping on mics, but let's do it for our own sake. More importantly, let's eliminate the distractions so that we can amplify the most important message people will ever hear.

Chapter 12 Discussion Questions:

1. Are distractions an acceptable part of your services every week? How could you eliminate them?

2. Do you think of your setting as "amateur" or "professional?"

3. What are some ways your team could help create an atmosphere for the congregation to have an encounter with God?

13

ASK THE QUESTIONS YOU NEED ANSWERS TO

When my wife and I were just starting to have kids, we took a parenting class. That's when we learned we had no idea what we were doing. It felt like the non-linear editing class I took before I had done any actual non-linear editing. It's useful information, but I had no clue what to pay attention to or what questions to ask until I'd done some actual non-linear editing. Until you have the experience, the class can only help you so much.

On the opposite extreme, as a technical artist, I have a pretty good idea of what needs to happen. I am the "expert" in doing live production at my church. Each week I work out the details of how to make services happen, and as a result, I know the answers I need from the creatives in order to succeed. However, unless I ask the right questions, I will not get the answers I need. I haven't always thought this way.

When I was starting out in production at my local church, I assumed that everyone thought the same way about services and that production was on the forefront of everyone's mind. How we would pull off some crazy idea was everyone's first thought, not just mine. I could not have been more wrong.

I used to get frustrated by this reality. I would do the classic tech-guy thing and become passive aggressive. I would shoot down every idea as impossible for one reason or another. I would get angry during set up or rehearsal when I found out key information that I should have been given days ago.

It finally dawned on me that most people I worked with were unaware of my needs. And not because they were self-absorbed, or they didn't care. They were wired differently and had their own issues and concerns to work out. They were asking different questions and looking for different answers.

If I was ever going to get the information I needed for the next service, I needed to ask the right questions.

Get the Answers You Need

When I didn't ask the right questions, I would get upset when the drummer showed up with that extra snare drum that I hadn't planned on. Or when we needed extra time to load in a new set piece. Or when we suddenly had a last-minute video to show.

If everyone is looking to me to have everything ready by the time rehearsal starts, I need to know as much information as I can get my hands on, and that only happens when I actually ask the right questions.

Clarify Concept Intent

Asking questions will result in a better understanding of what we are actually trying to do. I have been a part of so many misunderstandings caused by assuming I know, or they know, or that everyone is thinking the same thing. In reality, nobody is thinking the same way about any one idea. We are all coming at it from very different angles.

At the baseline, production exists to support an idea—someone's creativity. In our case, we are trying to create an environment where people can meet with God. Without fully understanding the intent of an idea, and how production can support this idea, things can get out of hand pretty quickly.

There was an era in my production life when I wasn't asking enough questions about intent. As a result, our production team just did what we thought was best, which more often than not turned out to be 180 degrees from the intent of the idea. Production can get pretty distracting when there is a misunderstanding about creative intent. If creating a distraction-free environment is a key value (and it should be), not understanding the point of an idea results in confusion.

Sometimes the right questions are about how things should feel. Not simply, "How many snare drums will the drummer bring this week?" But more like: "Do you want us to blast the audience with light, or do you want them in the dark at that moment in the song?" The kind of answers you are looking for are less black and white and more gray.

Getting answers to how production can support an idea can't happen unless you ask the right questions.

Start a Conversation

As you begin to ask questions for the sake of clarity, you'll start to see this go beyond just how to support an idea; you'll start to enhance the idea through the use of the technical arts. As your questions lead to you being more prepared and to making sure that everyone is on the same page with an idea, you will be building trust.

With the trust door opened, you can progress toward true collaboration, leveraging the technical arts to create something together. Elevating an idea beyond itself. This is a classic example of synergy—the sum of the parts is greater than the whole, and it can only happen if a true conversation happens between those who are creating the ideas and those who are executing them.

Chapter 13 Discussion Questions:

1. Have you been assuming that everyone knows what you need? How has this affected your team?

2. Think about some of your past event experiences. Would the right question at the right time have smoothed over some of the bumps?

3. What are some questions that need to get answered every week? Make them a part of your normal weekly routine.

14

PRODUCTION'S 80/20 RULE

I love to nail down the details. I want to know everything beforehand so I can be prepared. A technically flawless service is what I'm all about.

But, in my earlier days as a technical artist, I used to get frustrated when things kept changing at the last minute. Not necessarily because we hadn't planned enough, but because we were always trying to make things better. On the surface, I wanted to make things better, to always improve something so that it is as effective as possible. Few things got me riled up like deviating from the plan.

Didn't we have a production meeting about this? Why didn't we figure this out then? After years of banging my head against a wall, I started to think differently about what my expectations should be.

Many times, creativity needs to be seen and felt for us to know for sure if it is the right thing.

When I go to a concert or some other kind of show, I tend to be impressed by how well they have it all figured out. They show up and effortlessly blow the roof off the place. In reality, they've worked that show to death before I ever see it. They rehearse in a warehouse or off-off-off-Broadway until they make it the best it can be.

In our churches, we are doing something new each week and don't have the luxury of a test audience (unless you call the Saturday night service the test audience). No wonder it seems impossible to get everything right on paper before we show up to rehearse it. Many times, creativity needs to be seen and felt for us to know for sure if it is the right thing.

OK, so as the team that needs to execute this creativity, how do we handle the reality that we can't know exactly the way everything is going to go? Much of it is an expectation. When I'm in a production meeting, here's what I'm always thinking:

We can know about 80%, and 20% isn't knowable right now.

No matter how many questions we ask or how long we keep people in the meeting, we can only get at a certain percentage of what is actually going to happen. Knowing this helped me to relax a ton. If it is impossible to know everything, then that 80/20 rule feels about right. My job is to know and understand around 80% and to be OK with 20% floating out there that's unknowable.

We can know about 80%, and 20% isn't knowable right now.

This changed how I walked into a rehearsal. Instead of feeling like I had left no stone unturned and every detail figured out, only to be disappointed later when something changed, I knew there was a good chance that certain things would be fluid and changing.

Another positive outcome of not expecting to receive 100% accurate information in the production meeting was that I wasn't making the creative team feel like I was boxing them in. One of the worst things you can do to creative people is set up rigid perimeters for them to fit into. Production meetings can be the worst meetings ever for this group if they're surrounded by a bunch of problem solvers explaining why certain ideas won't work.

Once I realized that we could only know so much, I started approaching the creative team from the standpoint of wanting to know everything they knew. I don't need them to predict the future, but the goal is for us to be at the same starting line when we get to the weekend. What can we know right now, and what are we not exactly sure of? Can it be figured out now, or do we need to wait and see?

Once you know 80% of what's happening, it is usually a good idea to be aware of where the unknowable 20% is hiding. This way, when planning for rehearsal, you could add some extra time to figure it out in the moment.

Kill the 80%

The downside of the realization that you can't nail it all down in a meeting is that you have 20% of stuff in the unknowable category. "We'll figure it out on Saturday."

It takes a ton of discipline to keep at the 80%. There are so many things (around 80% of them) that can be figured out before you ever set foot in your auditorium or sanctuary, so we need to go after the 80% like crazy.

If you leave too many details to be figured out later, you'll run out of capacity to respond to the things that are truly unknowable.

Do Thursday's Work on Thursday

A key factor to being prepared to nail the 80% that you do know is that you've figured out what can only be done only on Saturday or Sunday when you are in the room. This list should be as small as possible.

Years ago, I used to meet a co-worker at the church offices at 4:30 a.m. From there we would head over to the high school where our volunteer set up team was ready to unload our three semi-trailers full of gear and get church up and running.

Inevitably, this co-worker would wait until 4:30 a.m. Sunday morning to print out and make copies of the input list and stage layout. Almost without fail the network would be down or the copy machine would be non-functional. He got mad at the IT department or the copier manufacturer.

We need to have margin for handling curveballs. We do this by going after the 80% we do know and destroying every last percentage point of it—before the weekend.

Maybe he was right and we needed more tech support for both the copier and printers in our office, but the reality is that we could have been getting these things done earlier in the week. We didn't need to wait until the weekend. Sunday mornings were already stressful enough without wondering if this was the week something wasn't going to print.

It should be our goal to make the "What I can get done on Thursday" list as long as possible so we're not leaving many tasks until game day.

Proofreading graphics? That can be done before the weekend.

Line check? That can be done before the weekend.

Figuring out how to stage a particular element? That can be done before the weekend.

Being fully ready for the 80% that you know, will give you the ability to handle the 20% as it presents itself in the moment. If you are only half ready, that means you have only prepared 40% and are leaving 60% to be figured out in the moment. This is a recipe for being frustrated and frantic, which will eventually wear down you and your teams.

We need to conserve our capacity for the unknown. We need to have margin for handling curveballs. We do this by going after the 80% we do know and destroying every last percentage point of it—before the weekend.

Chapter 14 Discussion Questions:

1. Think about the ways you are frustrated when changes happen. Are they caused by the expectation of knowing 100% before going into a rehearsal? How can you change your perspective on what it means to be prepared?

2. Sit down and write all the tasks that need to get done for a service to happen. How many can be done before the morning of? Do them.

3. What are some ways that collaboration would help know more of the 80%. Ask your creative team the best way to collaborate with them to achieve that.

15

PLAN B

Preparing for an event of any size requires coming up with a plan: input lists, building graphics, stage layouts for the volunteers to follow, etc. Developing a plan is why we have production meetings and send out emails to our team letting them know what to expect. Going after Plan A is what we are in the business of doing.

However, for Plan A to succeed, there needs to be a Plan B.

The Unknown

Lately, I've been having quite a few conversations about the idea of live production and how different it is from production done in a studio or any other controlled environment. Stuff happens in a live setting that is outside of our control, and Plan A isn't designed to handle those things. That doesn't mean we shouldn't work like crazy to make Plan A as solid as possible, but at a certain point Plan A can only take care of the things we can plan for. What about the unplanned-for things? Plan B.

Many years ago, I took my first real vacation and needed to find someone to replace me for two weekends. In those days, it was just my volunteer team and me. I had identified one person who would be my substitute for those weekends, so I compiled an exhaustive check list for him to follow, filled with every small detail that was in my head: things to check on, things to look out for, people to talk to. I left for vacation feeling great. Plan A was in place.

When I got back from vacation, I got an earful from the music director about some things that had gone wrong while I was out. At the time, we were a portable church and we were hauling gear all over the place. We had a pretty reliable system for getting stuff up and running quickly, and it generally functioned like clockwork. On one of the weekends I was gone, an equipment case that usually stayed on the trailer during the week had been taken off and as a result left behind the following weekend. This had never happened before.

The statement from the music director went something like: "THIS CAN NEVER HAPPEN AGAIN!!!!" (Yes, this really should be in all caps.) I think he also suggested that I should never take time off again. I remember thinking, "How can I prevent something from happening that has never happened before?"

It is impossible to plan for a situation that has never come up before. The only thing you can do moving forward is to develop a way to make sure the same thing doesn't happen again.

Mistakes are painful. Learning from them is easy. Use your past experiences to develop your Plan B. And make sure the people you work with also know that in order to avoid something similar from happening in the future, there is a new Plan B in place.

In these moments, learn to stay calm so you can think clearly and fix the problem. Worrying about Plan B for next time can happen later. For now, extend grace to yourself and your volunteers. Problem-solve now and figure out a better process when the dust settles.

Agree on Plan B

After digging through all that might happen and finally landing on the ideal plan, it is important that the team has a good sense of what we'll do if Plan A goes down in flames. In every situation, each person will react in a different way. Everyone will have their own idea of what to do, unless it is very clear what Plan B is going to be.

For big events like Christmas and Easter, we often use more technology than normal, which means we have more points of failure than normal. Once you start making things more complicated, there are plenty of opportunities for things to go wrong.

This is something I've learned the hard way. Because I've been burned so many times, I've made it a habit to have a Plan B meeting. If the video/Pro Tools sync goes down, what do we do? Do we run a backup tape? Do we create breaks in the video that we can go back to if it fails in the middle?

We've done some events that were so complicated, our Plan B was to reboot the computer and wait for it to get back online. While this doesn't sound like a great option, it is important to agree on what Plan B is before going into an event. Everyone needs to understand the plan, so when something goes down and we have to wait five long minutes for the computer to reboot, we all knew that it was a possibility.

If we haven't talked about it and agreed on a Plan B, chances are the expectation is that we simply flip a switch and everything is back up and running. Plan B can be almost anything, as long as everyone knows and understands what it is. Not only the production team, but also the worship leader and the pastors.

Everyday Plan B

Since each service can be slightly different, working through a potential Plan B in a meeting is key to developing a successful Plan A. However, there are some Plan B's that exist regardless of the specifics of each service.

There are a few things that always need to be included in Plan B. For example, wireless mics can fail, so have a couple of wired mics in the front row to hand up at a moment's notice. Have a locked-down wide shot to cut to on IMAG in case things get crazy. Put a general stage wash on a manual fader. Trust me, just do it.

These are all simple solutions that will keep the service moving forward. They aren't expensive Plan B's, and they aren't complicated. They are as simple and easy as possible for people to remember. For us, they have become second nature and things we don't even think twice about. At every event we do these things, just in case.

I was talking with a friend who runs a live video production company, and they have an XLR cable taped to all their camera fiber cables for intercom, even though intercom travels down the fiber. There have been enough times during events where the intercom has gone down on the camera, and this is their Plan B. Instead of waiting for it to happen, they pre-set Plan B and have it ready to go.

Understanding where your common failures occur is important in figuring out where your everyday Plan B's are hiding.

Not Plan C, D, E ...

Once you start thinking about what could go wrong, it can be easy to come up with a contingency plan for every possible scenario. You'll end up with too many solutions.

When we are in the moment and something stops working, we shouldn't have to get out a chart to figure out which plan we are going with. It should be something that we can all know and all execute.

If you have a video sending timecode to Pro Tools and it stops working, there shouldn't be a plan for each of the nine possible scenarios. We should eliminate as much troubleshooting in the moment as possible and just go to the back up. It needs to be simple.

In the heat of the moment, trying to agree if we are going with Plan C or D is the exact wrong place to be. We need to jump to the solution as soon as possible, which is Plan B.

I suppose this goes for Plan A also. Keeping things simple eliminates the possibility of failure. When something has tons of moving parts, we open ourselves up to the possibility those moving parts cease to move. I've lost track of how many times I've gone to the copy machine and something is jammed or broken. Considering how sophisticated they are, it isn't a huge surprise that it seems like it is broken more than it is working.

Are you creating a collating, stapling, whole punching, folding copy machine plan, or are your plans simpler, eliminating the potential for failure?

When you look at the complexity of your Plan A, where is the point of failure?

How Far Should Plan B Go?

My kids and I make fun of my wife for spending quite a bit of time developing worst-case scenarios for just about every situation. It can sometimes keep her up at night. In the world of live production, it is easy to get worked up about what might happen.

When you begin to think about all that could go wrong, Plan B's can get expensive. When you think about all the redundancy that exists for a regular NFL broadcast, you realize you can spend a lot of money on the back up: redundant power, an extra projector, another media server in standby mode, two of everything.

How much money should we invest in our backup plan? How much mental energy should we be expending on "what if" scenarios?

If we aren't careful, we can spend a majority of our capacity on whether or not something will fail. Pretty soon, trying to do our best is based on not screwing up instead of being free to do our best.

Plan B is an important part of Plan A's success. Use your history to develop a list of everyday Plan B's. Take advantage of your production meetings to figure out ways that things might fail and develop a back-up plan from there. Don't make Plan B so complicated that you are using the resources of time, money or mental capacity that could be used to make Plan A better.

Chapter 15 Discussion Questions:

1. List out your everyday Plan B's. Should there be more?

2. How could you restructure your production meetings to include figuring out Plan B's?

3. Examine your services. How many points of failure are there? Is it the right amount? Could you make it simpler?

16

EXCELLENCE v. PERFECTION

A co-worker and I recently discussed whether or not perfection was the goal of any production. If making things distraction-free is just another way of saying perfect, how can we avoid the idea of perfectionism?

Maybe I just have a problem with the word "perfect." What is it exactly? What does it apply to? How is it achieved? If we are talking about mics on and lights pointed at the right things and graphics being spelled correctly, then sure, let's make it perfect.

However, a lot of what we do in production can be subjective and the idea of "perfect" breaks down. What is the perfect mix? Perfect IMAG? What is the perfect service?

If perfection is the ultimate goal, how far are we willing to go to achieve it? How redundant are our systems? Do we run a generator during every service in case the power gets interrupted? Do we buy two of everything just in case? Should everyone know how everything works so everyone involved can know the answers to every possible question? Should we stay all night and rearrange the stage to make it perfect?

The number of things we could do to eliminate risk and ensure perfection would be a never-ending list, and most of us don't have that many resources: time, people and money.

Perfection As a Perception

Each year, to prepare for my role in the Global Leadership Summit, I would read as many of the books written by the presenters as possible. This helped me get my mind wrapped around the content of each session, and since I interacted with all the presenters, it helped me get to know them in some way before I met them. As a result, it was easier to make small talk about the topic they are passionate about while we are getting them mic'd up.

One year at the Leadership Summit, Dr. Brené Brown spoke about the science of shame and vulnerability. I have read her book "Daring Greatly," which I would highly recommend to every technical artist I know.

Here's a quote that I love:

> *Perfectionism is more about perception than internal motivation. Perception is impossible to control.*

As tech people, we tend to be accused of driving for perfectionism. Our attention to detail or our need to know exactly what is going to happen is because we want things to be perfect. That flawless execution is the highest value. Higher than if the element actually works or higher than if people are moved. As long as things go perfectly, the tech person is happy. The goal should be about a great service and not a perfect rehearsal.

For me and for the team I had the privilege to lead, we cared about flawless execution, but it wasn't because that's what matters most or because my self-worth is wrapped up in a service with no mistakes in it. The internal motivation is to remove potential distractions from people's experience. The goal is to make production as transparent (which sounds better than invisible) as possible so people are focused on worshiping or hearing the message without anything getting in the way.

The goal should be about a great service and not a perfect rehearsal.

I am pretty comfortable with the fact that I am not a perfectionist. I like to do my very best, which is all I can expect from myself. I am uncomfortable with being labeled a perfectionist, because that's someone else's perception of me.

While it might be impossible to control people's perceptions, as Brené states, I need to do my very best to try to change that perception. I can't expect people to understand my motivation or my team's motivation if I am not continually casting vision for why we do what we do. I'm not so worried about the random people who come to the tech booth and complain about the volume. I'm mostly interested in helping the people I work closely with understand why I am after so much information, or why rehearsing things matters so much to my team.

I can't expect people to understand my motivation or my team's motivation if I am not continually casting vision for why we do what we do.

This is where I really love the idea of excellence over perfection.

Excellence

"Doing the best with what you have." is one way to define excellence. This helps put things in perspective. You can only do your best, which sometimes might appear as perfection. This concept takes into account all the things that you've never experienced before, and it factors in the reality that stuff breaks. It considers the skills of your team and the type of equipment you have.

Another definition of excellence is "being better today than yesterday." This considers learning from mistakes and new experiences in order to keep getting better and better.

Excellence is something that can be attained. I have to work for it. It won't happen by itself, but I can go after it and get it. Excellence is within everyone's reach. It doesn't matter how much gear you have or how big your church is or how much your senior pastor values technology; you can do things with excellence. That's because it is all about doing the best with what you have. Some of us have more than others. Who cares? Be excellent with what is in front of you.

Doing the best with what you have.

In contrast, I can work until I die and never achieve perfection. It is not possible to work hard enough. It doesn't matter how much gear you have, or how big your budget is, or if your senior pastor knows what an SM58 looks like or not. You cannot work hard enough to achieve perfection. It is a treadmill that you cannot keep up with. Sooner rather than later, you will be thrown off the back of that thing and hurt yourself.

Being better today than yesterday.

When I am working toward a particular event, I am approaching it from the perspective of doing my very best. Using the resources of time, money and people, how can we achieve excellence? Finite resources force me to define what can and can't be done with excellence.

So, the goal isn't perfection, but rather doing your best and being better today than yesterday, which hopefully includes things being flawless ... most of the time.

Chapter 16 Discussion Questions:

1. What is your internal motivation for a flawless service?

2. How would you define the difference between perfectionism versus excellence?

3. What are some ways you could help change the perception people might have of perfectionism?

17

DEFINE NORMAL

As a tech person, I like figuring things out. Once I've figured something out, I also like to keep doing it the same way over and over again. Unfortunately, in the world of the local church, we are trying to come up with something new each week. That means the thing I figured out last week has been replaced by something completely new this week. This can be a pretty crazy roller coaster.

It reminds me of the movie "Jurassic Park." The first movie was jaw-dropping. I saw it during a preview night, and as soon as it ended, reporters were waiting at the doors to interview people—asking what they thought. I wondered what I would have said. All I could think of was stuff like "Two thumbs up!" or "Best movie of the summer!"

The first "Jurassic Park" movie required a ton of work to figure out how to make the dinosaurs seem real. I can only imagine the effort required by a handful of people to accomplish such a feat.

When the second "Jurassic Park" movie came out, it used basically the same technology as the first movie. Do you even remember the second movie? Whenever I wish that people would stop coming up with new creative ideas that I need to figure out or when I want to settle into the stuff I already know how to do, I remember the second Jurassic Park movie.

Not wanting to be part of a movie that's so forgettable, no one remembers its full name (Jurassic Park: The Lost World), I think most of the creative teams in the local church want to crank out Jurassic Park 1-level ideas. The reality is that the first "Jurassic Park ideas" take a lot more work than Jurassic "Park 2 ideas." As the person who has to get it done, it is easier to shoot down the really cool ideas because they seem to take too much time and money. So, what do we do?

One of the challenges I faced early on in my life as a church TD was estimating how much work something took. I knew I was working a little too much, and, as a general rule, I seemed to be running all over the place. I also knew that I was either saying yes to everything and wishing I hadn't, or I was saying no all the time and people found me difficult to work with (What?!?!?).

For the sake of my life's pace and for the relational damage I was doing by being difficult to work with (What?!?!?!), I needed to quantify the work that was being asked of me.

What Can Be Accomplished?

What I came to realize was that the people asking for things had no real idea how much time things took. On top of that, I had no tangible idea either. I just knew I was working hard and usually coming up short at the end.

I felt this the most when our church met in a high school theater, and we had one hour to set up before rehearsal. Without fail, we would start rehearsal late because the production team wasn't ready when the band showed up. I decided to start paying attention to what we could actually accomplish in that amount of time.

This exercise put some parameters around what could be done and what couldn't be done in one hour. When we saw it in writing, we were better able to make decisions about whether we could do a Jurassic Park 1 or a Jurassic Park 2 idea. We decided to call the one-hour set up "normal." We knew what we could get done in the "normal" amount of time.

If we had an idea that fell outside this boundary, we knew it, and we could come up with a plan to get it done. Did we need more volunteers that weekend? Should we ask to get into the school early? Should we not do something else so that the cool idea could happen?

Once we knew what we could pull off with the resources at our disposal, it made a huge difference in the conversations we could have. They turned away from me being defensive about how much time something would actually take and instead became a rational conversation about how to get it done.

From defining normal within that one hour, I moved onto defining normal for the rest of my week. What could I actually get done in a "normal" week? Let's pretend that a normal work week is 40 hours. What does that really mean? If the creative people have no clue how much time things take, how could I quantify the tasks that needed to get done in a week?

We definitely didn't hit normal every week, but just knowing what normal was supposed to be helped us make adjustments. Whether it was making an idea less complex, calling in more volunteers, saving the idea for another time, or renting equipment; knowing what normal was helped us understand when someone was asking for something beyond normal.

Priorities

Once you start keeping track of how much time things take and figuring out how you spend your time, you might notice that you are spending way too much time and energy on unimportant things and too little time on the things that matter. If your church is like mine, the weekend service is the most important thing. This is where the majority of my energy needs to go.

After going through this exercise, you might notice that the mid-week women's Bible study is getting an inordinate amount of your effort. Not to say that a woman's Bible study isn't important, but defining normal can help put the correct amount of effort toward it. Seeing where your time goes also helps your non-technical leaders see where the needs are.

Defining Normal After the Fact

It took me a few years to figure this out. When I started keeping track of how long things took and realizing that I was probably doing too much each week, I decided to put up more intense boundaries. The problem was that everyone had become used to my capacity and couldn't understand why one week I would say yes to them, and then suddenly I'm saying no to just about everything. And shutting down the women's Bible study wasn't an option.

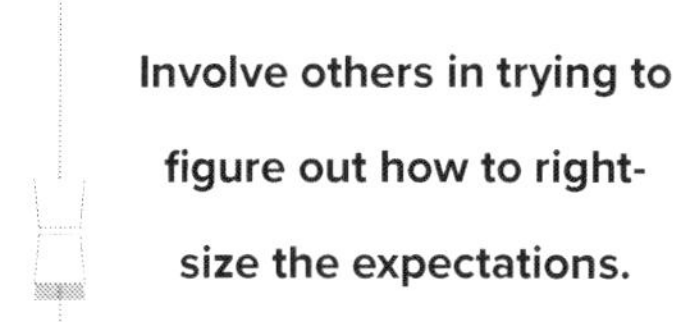

As we've established, most of what you do is already a mystery to the non-tech people around you. When you say you can't do as much, it can be difficult for them to understand why. Not only do you need to educate people on what it really takes to do production in the local church, but you most definitely need to ease into dialing things back to normal. If you just stop doing "normal" things cold turkey, you are going to upset quite a few people. I speak from experience.

Involve others in trying to figure out how to right-size the expectations. Talk with your boss or ministry leader to help define what can and can't be done. Bring your documentation to them so you have tangible, quantifiable data to show where the time is going.

This whole process of defining normal revolutionized my life. I went from feeling overworked and not really understanding why, to working hard and being able to prioritize where my energy went. Instead of always complaining that I had too much to do, I was able to brainstorm with my boss on ways to simplify an idea or rearrange my week to accommodate an idea.

Did this mean I never worked too many hours again? No. But this is a huge shift from being a passive, defensive tech person no one understands to being a healthy problem-solving team player.

Chapter 17 Discussion Questions:

1. What can you accomplish in a normal week? Write down a simple outline.

2. Over the next week, write down how you spend your time. Are there any places where you are spending too much time? Too little time?

3. What are some creative ways to reorient your time so that you can accomplish what needs to be done in a normal week?

18

WAIT FIVE MINUTES, THEN FREAK OUT

When I was the technical director of a church that met in a brand new, state-of-the-art high school theater, we had a pretty sweet setup. On the downside, we had three semi-trailers full of gear to set up and tear down each week. Fortunately, there was a great team of volunteers who knocked it out every week.

For the most part, we relied on a robust system for getting all this stuff set up and working on time. However, nothing always goes according to plan. Once, I overslept. More than once, the custodian who was supposed to open the building overslept. And I've lost count of how many times rooms in the building were double-booked.

I can remember one such time when the cafeteria, our video overflow room, was also rented out to some kind of yoga group. When we realized there was an issue, most of my team started to freak out. Not just a mild freak-out, but a full on "how can church possibly still happen" and "this is a major disaster" type of panic.

In that moment, I wanted to join in with the whole team freak-out. Maybe it could have been a great bonding moment for us. But instead, I remember thinking that while I wanted to panic, I was pretty sure that wouldn't help any of us. Not only would it just be one more person losing it, but as the leader, I was certain my team needed to see me stay calm.

In the split second it took my brain to go through this exercise, I decided to stay calm for five minutes while I tried to solve the problem. After five minutes, if we hadn't come up with a solution, I would join my team in freaking out. But for five minutes I would hold it together for the sake of my team.

For five minutes I would hold it together for the sake of my team.

In this example, and in pretty much every example since I started living with this mantra, we would always uncover a solution within the five-minute window.

The Unexpected

Things come up that we haven't planned for and can't plan for. Something is broken, a volunteer doesn't show, the building is locked. You get the idea. How do you handle yourself in these situations? How can you manage the stress in those moments?

When you are confronted by a tough situation, what is your knee-jerk reaction? If you are the leader of a team, or even if you are a member of a team, how you respond in these moments will determine how easy it is to follow you or how reliably you handle yourself in a crisis.

Not only is it bad for your immediate team to see you lose it, but all the non-tech people you work with will definitely question your ability to get stuff done. On a good day, they don't know what's going on. When they see you in a full-on panic, they are going to wonder if you're up to the task.

When I was in high school, I was the kid in the booth at church. I loved doing production, but I didn't necessarily think I was any better than other people on the team. It turns out that I wasn't technically better, but I was able to react calmly when something bad happened in the moment. After coming back from a weekend off, people would breathe a sigh of relief: Todd is back! Now nothing bad will happen! In reality, the problems that existed on the weeks I wasn't there were the exact same issues we had when I was there. How I responded was the difference.

Staying calm in a crisis can make all the difference. Not only for your team but for those who are in the dark about how production works in the first place.

Everyday Stuff

I used to work with someone who ran everywhere. It didn't matter what was going on or how well it was going, this person ran all the time. I guess he just wanted to get stuff done faster. The first time I experienced this, I thought to myself, "Dude, stop running! You're making me anxious." I could look around at the team and see that others felt a similar way.

In reality, he was not prepared for the normal work of doing a weekend production. He had not done Thursday's tasks on Thursday and had left too much to be done on Sunday morning. As a result, the only way to get it all done was to run from one task to the next.

Any time you see a leader who seems rushed, you can almost guarantee that their team will seem rushed, whether or not they actually are. In this example, much of the team was standing around watching this person run around. They had the capacity to help, but weren't able to because the leader had taken on too much.

What Your Response Communicates

As a leader, how you respond in a given situation inspires something in your team and in the people you work for. It can either be confidence or panic. Each response helps build into one or the other, and when an emergency happens, what you've built up will work for you or against you.

One year, right before the start of Session 1 at The Global Leadership Summit, we had a hiccup in the power at our facility. It knocked out our lighting console. If ever there was a moment to panic, it was now! From my vantage point down near the stage, I looked back and saw the team running around trying to reboot the lighting system.

What I felt inside was the urge to run back there and find out what was happening. Then I thought if I start asking too many questions, I would be pulling them away from actually solving the problem, and the people back there trying to solve the problem were our very best people to have on it.

When my leaders asked what was happening, I calmly responded with:

"Our best people are back there working on it."

This response communicated a couple of things. It communicated to my leadership that I have confidence in my people, and it communicated to my people that I have confidence in them. To run back to the booth and start yelling at people to fix the problem communicated exactly the opposite to each group. Whether or not the problem gets solved in the best way possible, by not panicking, you are communicating that everything will be OK. For techs and non-techs alike, that is an important thing to realize.

When I think back on that experience, I still get a knot in my stomach similar to the one I had while it was going on. I'm not saying it is easy to hold it all together, but the benefits of waiting five minutes and then freaking out, pay off 99% of the time.

Since there is already enough stress around setups, rehearsals and services, why not take a deep breath and wait five minutes before freaking out?

Chapter 18 Discussion Questions:

1. How would you characterize the vibe of your production environment? Is it panicky or calm?

2. Think about a recent crisis situation. What is the worst that could have happened by waiting five minutes before panicking?

3. How does your church leadership view your team in times of crisis? How could you work to improve that?

19

PACE YOURSELF

I love the big idea. I love going to concerts and seeing something incredible. I do my best to watch most of the award shows, because I know there is a good chance I will be inspired to try something new. Eurovision, anyone?

As tech people, we get excited about doing something cool. Whether it's using new gear or using old gear in a new way, we love pulling off an element that people think is awesome. But sometimes we tend to get more pumped about some cool tech thing than about the effect it will have on people's lives. That's a topic for another time.

One of the challenges of trying something new is that you have no idea what you're in for. You may think you do, but you don't. I've done quite a few home improvement projects. The first time I finished a basement, it took ten times longer than I thought it would, and without fail, every time I've tackled some new aspect of home improvement, the 10:1 rule has applied. Doing something for the first time always costs more and takes more time than I'd originally thought.

In the land of church production, the same idea applies. With the challenge of having a new service once every seven days, it is easy to see why trying to come up with something amazing every week can get exhausting. And I would say it's not healthy or possible long term.

So, what do we do? How do we balance keeping things fresh with having something sustainable? Without knowing what you can expect from yourself each week, it is really hard to know how much capacity you have for pushing yourself.

The other part of this is to pace yourself.

Unfortunately, at this point, it is time for the "running a marathon" analogy. I've never run that far, but I have run a half-marathon. When you are used to running three miles, it is easy to start your 13.1 mile run at the pace you are used to running. What I found was that I couldn't keep it up. By mile 12, I could barely function. If I had run each mile at my normal pace, I would have had to drop out of the race.

Pacing yourself while running is not easy. Pacing yourself in life is no cake walk either. There isn't a formula that works for everyone. Much of this will depend on your own personal capacity and that of your team. Even though you might be capable of doing vast amounts of work, you will still have limits that you will bump into. Each of us needs to figure out what pace we should run.

In most every church setting, we have a couple of really big moments in the year where we need to push and try something new: Christmas and Easter. Beyond that, it is up to us to figure out how often we should stretch for something just out of reach. There needs to be some kind of rhythm, or else you will be pushing yourself constantly, which is too often.

Recovery

When we push the limits of our abilities every week, without time for recovery, we will find that kind of life is unsustainable.

If you are constantly pushing yourself with an exercise routine without a chance for your body to recover, you will start doing more harm than good. Your body can't keep up the pace.

In the world of production, too many long nights and too many crazy deadlines will begin to take their toll. Pretty soon you will become less and less effective. Your ability to make good decisions will become clouded by how weary you are.

I believe this is where many tech people get trapped. OK, if I am more specific, this is where I get trapped. If I'm just doing whatever someone is asking, and if I haven't learned the art of pacing myself, pretty soon I'm grumpy and feeling like nobody cares about me. I lived a good part of my early years this way.

Be honest. Frustrated and grumpy describes the stereotypical tech person in the local church. Much of this comes from an unsustainable pace. It is easy to become the victim in this situation, but much of it comes down to our own willingness to admit we can run only so fast, and that there needs to be time to recover.

When I first started in ministry, I loved it and worked seven days a week. This was before I got married and learned my fiancé wasn't a huge fan of this kind of work schedule. After we married, she decided to go talk to the senior pastor about how they were working me too hard. After he listened to what she had to say, he responded with "Nobody is telling him to work that hard. If he needs to pace himself, then he needs to be a man and do it." Ouch.

I believe that many of us who feel like we are working too hard are doing it to ourselves. Sure, the church places demands on us, but if we aren't standing up for what we can sustain, who is?

Taking time to recover from Christmas should be just as important as the process leading up to Christmas. It's just as important to catch your breath as it is to come up with an amazing idea. And trust me, if the pace of your life is not sustainable, you won't have too many Christmases to dream about.

On the other hand, if your pace is not fast enough, you've got trouble ahead. If we're only worried about making things easy and doable, you'll become stagnant.

At a certain point we need to stretch ourselves.

Stretch Yourself for Inspiration

Stretching ourselves out of our comfort zone, but not to the breaking point, keeps us engaged. More than that, it helps to inspire us for the next big thing. With the regularity of services, it can be easy to get worn down by the grind of the same thing week after week. By thinking outside the box and then pushing to accomplish something new, we are pulling ourselves out of the daily routine and expanding our horizons.

Trying something that you've never done before kicks another part of your brain and your heart into action. To problem-solve, to dream, to imagine what could be—these are all things that don't necessarily happen on a normal weekend.

When my team reminisces about the large events, we always end up remembering the times when we really pushed ourselves. We talk about the times when we did incredible things together and the times when it was a disaster. But the moments in our past that inspired us are generally the ones when we went for it. Honestly, anything that I see or hear about that inspires me from a live-production standpoint, happens when someone has stretched beyond what they have normally done. Whether it's a great set, or some cool new environmental projection, the new and different can be very inspiring.

Learn how to pace yourself and your team so that you can be inspired by the moments of stretch instead of worn down by them.

Stretch Yourself to Increase Capacity

Pushing ourselves increases our capacity. Going back to the running analogy, I know if I increase my pace a little at certain intervals, pretty soon my whole pace has increased.

In the world of production, once you try something for the first time, it has the potential to be a part of your normal bag of tricks. The thing that almost killed you two Christmases ago is now something you don't hesitate to do on a normal weekend. Pushing yourself to try something new at the right interval means that you are increasing your endurance, your capacity, your normal.

Without pushing yourself from time to time, you will stagnate. Pushing yourself all the time will run you into the ground, which means the race is over. Learning the proper balance between the two is key.

Chapter 19 Discussion Questions:

1. Is the pace you are currently running sustainable? Why or why not?

2. Are you bored with the same thing every week?

3. How could you inspire yourself or your team to stretch?

20

GOOD PRODUCTION ISN'T EVERYTHING

Like me, most tech people I know think only about production. It's not that I don't care about the content, or if it is what the church needs to hear in that exact moment. But my main concern is: Does it all sound and look good, and can I get my job done with what you are giving me?

Technical artists in the local church can suffer from tunnel vision when it comes to the task at hand. And it's no wonder; there's a lot to do! I suppose it makes sense, because it is how God designed us to work. He created the technical artist to care about the technical arts. Someone else is created to think about all that other stuff.

What happens when the best choice for production isn't necessarily what's best for the service or best for the church? I know it's difficult to believe that production isn't the most important thing going, but it happens. It probably happens more often than not.

Care Anyway

On one hand, it is easy to care about the things that matter most to me. In reality, it can get exhausting pushing against all the other needs of our church stacked up against what matters to production. Since you are probably the only one fighting for production, it is easy to become discouraged. Care anyway.

In 1 Corinthians 12, when Paul talks about all the different parts that make up the body of Christ, it's clear that without each of them, the body doesn't function properly. If you don't care deeply about your part, the body doesn't function properly. If you don't figure out how much rehearsal time your team needs, or how many inputs you can handle this weekend, or figuring a replacement plan for the old gear, the church will not reach its fullest potential.

The flip side of 1 Corinthians 12 is that the body of Christ doesn't function properly without the other people caring deeply about the things that God created them to care about. This can cause tension, but as we push and pull against each other, the body gets stronger. As we learn how to work together, the body of Christ reaches greater functionality, but this can only happen when you don't give up, and you continue to care about production.

Now Let Go

As a technical artist, I find myself facilitating someone else's ideas much of the time. Whether it is the worship leader, or the youth director or the senior pastor, we all have people who are making bigger decisions than the ones we're making.

The kinds of decisions I have to make involve knowing which mic to use, how many lights to turn on and what kind of font to use for graphics. Not that these decisions are always the easiest, but production decisions follow other, larger decisions that affect the whole church. Sometimes those decisions aren't in the best interest of our production process. It could compromise the mix, or the process, or is riskier because we haven't rehearsed it yet.

How do you care deeply about something, go after it with determination and then let it go?

When it comes down to what is best for the service or the church, what matters to you isn't always the most important. How do you care deeply about something, go after it with determination and then let it go?

A big part of it is to be comfortable with what might happen next. If you know that all the information is on the table, and that the leader has everything she needs to make a decision, then you need to go with her decision. And don't respond with your passive aggressive self but with the same enthusiasm with which you'd go after your own ideas.

To be a real team player, it is important for your leaders to know that you will give them your best, whether you get your way or not.

If your concerns are realized, it isn't your fault since you did such a masterful job of informing your leader of what could happen. You still did your very best, even though it wasn't your recommendation.

To be a real team player, it is important for your leaders to know that you will give them your best, whether you get your way or not. They also need to know that you will take care of your stuff with full ownership so that they never have to worry about it.

Trust is built by accumulating encounters like this every day. Trust that you are taking care of production and trust that the best overall decisions are being made. And trust is the foundation of the body of Christ when it is functioning at its fullest.

Chapter 20 Discussion Questions:

1. Are you getting worn down by the amount of effort it takes to push the needs of production at your church?

2. When your leader does the opposite of your recommendation, what is your first thought?

3. Does the phrase "I told you this would happen" ever come out of your mouth? How can you constructively talk about what didn't work?

21

CRITICAL PATH

In college I studied Industrial Engineering. How I got there is a long story, but once I had landed on IE as my major, I really enjoyed it. When I graduated, I was the only one in my class with a job. Granted, it was at a church doing production stuff, and there were fewer than 10 people graduating that quarter, but it was a job, people.

From the outside, this church job had nothing to do with Industrial Engineering. But the reality was that the whole thing was like a real-life IE story with real-life IE problems. How do we set up two semi-trailers worth of production gear and children's ministry gack in under two hours, with volunteers who have no prior experience, then put it all away at the end of the day?

Those early days were a lot of work, but the training I received in Industrial Engineering really paid off. That and the fact that the person who led this process before me was also an Industrial Engineer.

Critical Path: the order in which a series of interdependent operations should be done so that a project can be finished as quickly as possible.

One of the Industrial Engineering concepts that had the most impact on my role as a TD in the local church was the concept of the critical path.

Looking around the internet, the varying definitions of "the critical path" can make your head spin, but here's the one I liked the best for our purposes here:

The order in which a series of interdependent operations should be done so that a project can be finished as quickly as possible.

When there is only so much time in a week or on a Sunday morning to get everything set up, how can we maximize the time we have with the tasks that need to get done?

I don't know about you, but I can be easily distracted by wanting to spend time on things that are fun or interesting but may not be the most important. Time critical tasks need to get done first.

Make a List of Everything

Figuring out the fastest way to get things done means you need to start by making a list of everything that needs to get done. To find the critical path, you must know the totality of things that need to be accomplished. Not only should you know the tasks themselves, but you need to assign an appropriate amount of time to each one.

Once you have this list compiled, it is time to start prioritizing them. What needs to be tackled first? Some items on the list can't be done until other tasks are completed. Some of the other tasks can be done any time. Putting this order together will help determine what needs to be done first, second, third, etc., so that you can accomplish the task at hand in the fastest, most efficient way possible. Stage set up can't happen before the truck is unloaded. Mic check needs to happen after the console is set up. Making sure the parent notification system works can happen any time before the start of the service.

Going back a few years, we were in a rented facility and getting ready for our Christmas services. We had probably bitten off more than we could chew as far as the set and program were concerned. (That's what Christmas is about, right?) We naturally had too much to get done in a short amount of time.

As the rehearsal start-time approached, I noticed that my key audio person was installing a mechanism to make it snow outside one of the windows on the set. It was definitely something that needed to get done, but we still had hours. It didn't really have to work until right before we opened the doors. However, for sound check to start, we needed to finish wiring the stage, which we hadn't done.

I couldn't believe it! Are you kidding me? Put down the drill and get to the soundboard! Almost anybody else can set up the snow thing, but nobody else can make sound check happen!

In any production big or small, there are many decisions to be made and many, many, MANY tasks that need to get done. It is so easy to get buried by all you have to do, so much so that you lose sight of the most important thing. You need to figure out which tasks are time sensitive and dependent on other tasks and which ones stand alone or can be put off until the last minute.

Once you string together the time-sensitive and dependent tasks, you have your critical path—the shortest distance to the finish line. In a world where every second matters, your team and your church are counting on you to figure out what this looks like before you're buried by other non-critical tasks.

A Wrench in the Works

Without fail, once we are in a situation when time is of the essence, someone throws in something extra and unplanned for. It's nothing new, but now that you are thinking about your critical path, this minor inconvenience can trip up all kinds of other dominoes in the system. What might have just been a feeling before is now something measurable. If we take time to add an extra instrument or two, we are going to extend our setup time, which will push back rehearsal start time, which will affect the amount of time we have to actually rehearse.

When you've defined how much time a task takes and where it fits into the whole puzzle, it is much easier to see where things are getting pushed behind. Instead of having a vague idea of what is causing rehearsal to be late, you have tangible evidence that shows what needs to be adjusted.

As new things are thrown your way at the last minute, you will be able to communicate clearly the impact any changes will make to the whole. By spending time creating that critical path, documenting how long certain tasks take and how they all interrelate, your team and your leaders will see that you have a carefully crafted plan. When that plan gets altered by someone else's poor planning, the burden of responsibility for things falling behind schedule falls on them, not you.

I used to get frustrated when someone added things at the last minute, and people asked me why we started rehearsal late. Once I had figured out the critical path, I could preempt the questions by informing people that making a last-minute change would put us behind and could estimate by how long. If that was OK, we'd do it. By talking about it before we actually made the change, I was putting the responsibility of the late start on the person who should feel the responsibility instead of absorbing it myself.

When you have a critical path plan, you are also taking the information out of your head and sharing it with those on your team. By having a written plan, your team can help carry the heavy load of getting tasks done. With a written plan, each person can understand how the tasks are interconnected and what the shortest distance to the finish line looks like.

Chapter 21 Discussion Questions:

1. Create a critical path for a rehearsal. Write out a list of everything that needs to be done to get ready for rehearsal. Now put them in the order they need to be done. Which are dependent on each other? Which aren't related to anything else and can be done whenever?

2. When things are added at the last minute, do you automatically do them without question? How can you share responsibility for this decision with the person making the last-minute request?

3. Does everyone on the team know what the critical steps are to achieve success?

22

PRODUCTION IS TOUGH. LET'S ENJOY IT

Recently, I was leading a team that handled the production needs for a concert tour. For those of you who have done something like this, you know it is a ton of work. There was someone on the team who wasn't really a production person and who didn't normally serve with us. He kept saying how much work production was and that normal people didn't have a clue what it took to pull off what we do on a normal basis.

I agree with him. Production is tough stuff. There is a ton of work that needs to get done. Lots of physical labor. Lots of details to manage. A crew of volunteers to keep track of and keep moving forward.

I've been doing production work for a long time, and it never seems to get easier. When we build in new systems or buy new gear to streamline the process, we usually increase our capacity and then push ourselves to that new standard.

I often tell myself that what I am privileged to do is not easy. But if I'm going to spend my life doing it, I want to enjoy it along the way.

Don't Take Yourself Too Seriously

I worked at a place where the phrase "every week is like the Super Bowl" was said often. From a production standpoint, who can do the Super Bowl every week? (Who wants to?)

In reality, there is no way to keep up this level of intensity. It is so easy to get wrapped up in not making a mistake that you end up creating more pressure on yourself and your team than actually exists.

This somewhat artificial pressure can steal the joy of serving from you and your volunteers. I don't know about you, but joyless serving doesn't sound like God's plan.

I love that God has chosen to use us and our gifts to help accomplish his purposes on this planet. I don't understand it, but I'm grateful that I get to be a part of bringing the kingdom of heaven to earth. However, if I am so worked up about not making a mistake, am I really contributing to bringing heaven to earth? Isn't it true that God is infinite and is able to do things that are beyond my comprehension? Couldn't he work in people's lives whether or not there was audio feedback?

This is a giant mystery to me. God has chosen to use us, but he doesn't need us. We have to do everything we can to create an environment where people can meet with God in our services, but at the end of the day, God doesn't need us in order to meet with his people.

So, with this as the backdrop, let's stop taking ourselves too seriously. Yes, we have lots of work to do. Yes, it is important. Relax. Enjoy yourself.

Make Space for Fun

There isn't always time to have fun—especially for those of us who have an early load-in on Sunday morning. It's not like we have much opportunity to sit in a circle and sing Kumbaya. We've got work to do!

How can we be open to enjoying the process of getting things set up and ready? How can we make space for having fun with each other as we go?

If I look back on my years of serving in production, they are jam-packed with fun moments we remember, even after so many years have gone by. These "fun" moments were mostly in passing, while something else was happening. Whether it is an inside joke or something amusing that happened while we were tearing down, recognizing and remembering these moments is important.

Right now, I'm thinking of all the times we had "road case races" down the hallway to help speed up tear down. Or when we pulled out a football to throw around the auditorium. Or when we used to hide something on the stage during a service to see if anyone would notice. These are all examples of ways to have fun that don't take up any time but can help make the time we are together more enjoyable.

I'm not suggesting that we jeopardize our programs for the sake of a joke. We have jobs to do and letting fun get in the way of the task at hand is not what your church has in mind.

Save the Show

At Willow Creek, we had a tradition called "Save the Show." We would all go out after a full day and night of rehearsals. Even though we'd already spent a lot of time together, we wanted (and needed) to connect outside of the huge tasks in front of us.

"Save the Show" usually occurred around Christmas time, when we were working a ton already. But for many of us, by the time we were done with rehearsal, our families were already in bed and wouldn't miss us if we stayed out for another hour or so.

For our team, this was a valuable time to have fun and enjoy being together. When you are in the pressure cooker of a rehearsal, it is nice to hit the steam-release button together; to talk about something other than the problems we had to solve tomorrow. Relationships need more than one dimension. To work effectively with your team, it is important to have more than one aspect to build upon.

This "Save the Show" idea can also be extended to other opportunities for you and your team to be together. While you might be slammed during your actual rehearsal, it is important to figure out other times to get together—a celebration dinner after the huge Christmas run, or a field trip to a production company or going to a concert. Any time you can break away from the normal routine and have fun outside of work, you will bank more relational equity for the next time you're neck deep in a task together.

Production is tough. Let's do what we can to enjoy doing it together.

Chapter 22 Discussion Questions:

1. Has every week turned into the Super Bowl? How can you ratchet down the intensity?

2. What are some ways you could build fun, non-production tasks into your regular routine?

3. When was the last time you did something with your team outside of the tasks? What is one thing you could do moving forward?

PART 3

COLLABORATION

23

TENACITY IN RELATIONSHIPS

I called an old friend for advice on some challenges I was having with my worship leader counterpart. During the conversation, I asked how she and her worship leader got along so well. Her answer didn't help me. In her words, they had been through hell together, which ended up making their relationship stronger, which in turn made working together even better.

I realize in order to have strong levels of collaboration with someone, you need to have quite a bit of trust. And trust isn't something that happens overnight. It happens in small ways every day. Trust happens in production meetings and rehearsals. Trust is built in the edit suite and during the crazy hairball moment in the service. And trust is a two-way street. Our actions tell a story. Are your actions and your choices worthy of someone's trust? Are you choosing to trust the other person or are you always assuming the worst of someone else?

For us to move beyond the basics of production and begin creating and collaborating with the creative arts team, there needs to be loads of trust. Just as it requires large amounts of tenacity to nail the basics of production, it takes tenacity to keep going after the collaborative spirit. It cannot and will not happen by itself, and if it isn't worked on, it will decrease. It is like a white painted fence. If you don't keep painting it, eventually, it will no longer be white.

Our actions tell a story. Are your actions and your choices worthy of someone's trust?

Let's look quickly at the two halves of trust.

First, there is the trust that comes from doing your thing well.

Do Your Part

Have you been tenacious in the basics? Is production firing on all cylinders? Are you executing the service with excellence? Have you learned how to pace yourself? Are you prepared when rehearsal starts? All these things speak to how much you can be trusted to do your part. These are the values that you hold to, the things nobody has any idea need to happen to make the service work. They just need to be done. Are you doing them?

Once you've laid this foundation of trust, it is time for the second part: extend trust to those with whom you partner. This is probably the most difficult part of being a technical artist in the local church, and it's the secret nobody tells you about.

Experienced technical artists soon realize that doing great production work is the easy part. After all the time we've spent making the foundation of production everything it can be, no wonder we are reluctant to put it into the hands of someone else.

Trust in Others

As we look toward the fusion between the creative and the technical arts, putting your trust in others is the key to increasing the levels at which your teams will work together. The trust others have in you and the trust that you place in others will determine how well you collaborate and the capacity your teams have for God to use them.

Much of what we are going to talk about in this next section will center around ways in which you can begin to place yourself and your team in the right posture to collaborate with creative ideas. We will look at ideas for fostering the relationship between you and the creative artists.

While trust is a two-way street, you only control about 50% of the equation. You can't make someone respond a certain way or care about something as much as you would like. But you can help build trust by extending it where you can and taking control of the parts you do have control over.

I've already said this, but it is not easy. Trust requires patience and raw tenacity. Without it, your church will fall short of its full potential.

As we've learned, God created the body of Christ to work in community to help bring the kingdom of God to earth. And in case you haven't noticed, God made technical and creative artists very different from each other. For whatever reason, this is God's plan. Not the one I would choose, but nonetheless, it's the plan we have to work with.

In the following chapters, we will look at the concepts that you have control over. The preceding chapters have been about the tenacity required to nail the basics of production. In the next section, we'll look at being tenacious in the ways we work with our creative arts counterparts.

Chapter 23 Discussion Questions:

1. What is more difficult for you, developing the production foundation or collaborating with your creative counterparts?

2. What is the current level of trust between the technical and creative arts?

3. What is one action step you can take to strengthen the trust between you and your counterparts?

24

TECH PEOPLE ARE FROM MARS; EVERYONE ELSE IS FROM VENUS

One year, the arts team staff decided we should all go see a show together. We looked through the paper for something new and cool coming to town. That particular year, the show "Blast" was touring the country and also being shown on PBS. It was essentially a drum and bugle corps. Those of us who were band geeks in high school, including my boss who had been my band director from the 6th grade on, decided it would be a great experience for us to be inspired together.

The show turned out to be super creative, with great lighting, and performed with excellence. The set was a cool Hollywood Squares-type of grid that had different percussion instruments inside each one. After the show was over, we went out to eat together and couldn't stop talking about how cool it was. Most of my thoughts were about how they had done certain parts of the production: Mic'ing different instruments, lighting certain parts of the set, the kind of hazers they used. Sure, it was inspiring, but it also seemed unattainable. Who can afford that many wireless mics? How would we even build a set like that? They have more black lights than we have regular lights!

The arts director, on the other hand, was inspired to think of how we could somehow duplicate what we had seen. He began to brainstorm ways we could adapt that show into something cool in our context. Naturally, I started to explain to him why we couldn't do any of it. Technically, it was way past anything we had ever done and, like I said, seemed unattainable.

This show was sometime in October and despite all my warnings, you guessed it, we did it for Christmas that year—Hollywood Squares set and all. After that, we jokingly made a pact that we would never let our boss go and see another show. We knew that whatever we saw, we would have to duplicate later. Maybe we could let him go to a high school production, but that was about it.

The Tension

Sometime after this event, I began to think about how different I was from my boss. Given any issue, we would come at it differently, and usually from opposite sides. "Blast" sparked his creativity. "How could we do something like this?" When I saw "Blast," I tried to figure it out: "How did they do that?" which was followed by "We could never do that."

This difference between us used to drive me crazy. The arts director was always asking for something that seemed impossible, and I assumed he must know he was asking the impossible. I used to just say yes, thinking he already knew the technical ramifications of what he was asking and that we would probably do it regardless of anything I might have to say about it.

After years of carrying this attitude, I became angrier, and like so many other technical people, passive aggressive. I let my anger simmer inside to erupt at some undisclosed time. I would become the victim in every situation, grudgingly saying yes to all kinds of ideas and letting everyone know that I was not happy about the decisions we were making.

It finally dawned on me that I might be all wrong in my thinking. Maybe there is a reason that I think differently from everyone else. Maybe there is a reason my arts director thinks differently. Is there something to this? I began to read up on how God had designed the Body of Christ to function in an attempt to understand this dynamic. 1 Corinthians 12:14 says:

> *Now the body is not made up of one part but of many.*

Could this mean that my role is different from everybody else's? And maybe my arts director's role is different from mine? I have been created specifically to bring my unique gifts to the table. No one else had these same gifts, including my arts director. My arts director had also been created specifically to bring his unique gifts to the table. No one else had his gifts, including me.

This was a revelation to me! I always knew that we were very different, but the realities of that never occurred to me. He would NEVER think like me. The technical ramifications of his ideas would be the last thing he thought of. Maybe he wouldn't think of them at all. Similarly, I would never think like him. God created him to be creative, to think big, to think outside of the technical. God created me to think of the details, to think based on the current resources, to think inside the technical.

I have been created specifically to bring my unique gifts to the table.

Check this out from 1 Corinthians 12:18-20:

> *But in fact God has arranged the parts in the body, every one of them, just as he wanted them to be. If they were all one part, where would the body be? As it is, there are many parts, but one body. (1 Cor. 12:18-20)*

This tension I was feeling was part of God's plan. He designed us to be different from each other. In order for us to accomplish the most, we needed to bring our unique gifts together. We're supposed to be different. Nobody else thinks of the technical issues like I do, and that's the way God designed it. My arts director is always asking for the impossible, and that's the way God designed it.

This concept was a huge relief to me. I could breathe easy. I was the only one who thought like me, so I stopped assuming that everyone knew the technical issues associated with any idea. They wouldn't know unless I told them. I could stop getting frustrated or grudgingly saying yes to everything. In a normal conversation, I could discuss the technical issues of a creative idea.

If I didn't bring that to the table, how could I complain that everybody was running over me or that I was a victim of the creative process? What I have to say IS a part of the creative process. Without my contribution, we are not all that God wants us to be.

The unfortunate part of this "we're supposed to be different" thing is that it is tough. Because we all come at the same problem from completely different points of view, the chances that we disagree are huge. The chances that we misunderstand one another are huge. The chances we get your feelings hurt are huge. Playing your role in the body of Christ has the potential to be the hardest thing you have ever done. It also has the potential to be one of the most fulfilling experiences of your life.

If it were up to me, I wouldn't have created the Body of Christ this way. Whose crazy idea was it to put people who are so different in such close proximity to each other? I would have everyone more alike, which would increase the possibilities of getting along. For whatever reason, God decided that His plan was the perfect arrangement for His Body. Put the creative people right next to the detail/execution people. This will have the greatest kingdom impact. Since this is obviously the reality, there must be a way for us to thrive together.

Understanding that you bring something unique and indispensable to the body of Christ is the principle to hang your hat on. Understanding that others also bring something unique and indispensable to the Body of Christ is equally critical to remember. When things get tough, remember that the church will not function properly unless you can learn to work with the people who have been placed around you.

More Tension

When I graduated from college, most people thought I would be wasting my degree in Industrial Engineering by working at a church. Little did I know that when I started at Kensington it would basically be an Industrial Engineering story problem:

> A church of 800 adults meets once a week in a rented facility. You are responsible for transporting and setting up the equipment necessary to accomplish a full program including, but not limited to, audio equipment for a 5-piece rock band, and all the necessary items to provide an excellent children's ministry in about 15 separate rooms. You have two hours to complete. Can it be done? Explain.

The answer, in a nutshell, is make it efficient. For years, my whole world revolved around making transportation, set up and tear down as efficient as possible. I had to figure out what the constraints were, what the variables were, what the common denominators were, how to make a task so easy a first-time volunteer could do it, and complete it all in two hours or less.

Being involved in production means that part of our artistry comes from making it efficient. We view much of our world in terms of being efficient. Standardize a process so we can learn how to do it and repeat it with consistency. Create deadlines so we have time to complete the task with equal excellence every time. We ask for a 100%-accurate service order 10 days in advance so we can have the best chance of executing with excellence.

In God's perfect plan, he created us this way. He also created creative idea people. "Artistic" people. Musicians. They don't think in terms of efficiency. They think in terms of possibility.

While we want to streamline the process, creative ideas come from a very different place. While we want to know exactly what is going to happen, creative ideas are hard to nail down. Creative people always want to do new things, which means we haven't figured out how to do it yet. It also means it will take more time than if we just did it the same as last week. Creativity is inefficient in almost every way—the opposite of how the technical mind works. If we were to switch places with the "creative" people, none of the services would be set up on time because they would be trying new and exciting ways of doing set up. If we were responsible for planning the services, they would be predictable and boring because we'd figure out how to do it one way and institute a "wash-rinse-repeat" system.

As I got more involved in watercolor painting with my daughter, it became increasingly maddening. As a person who thinks in terms of efficiency, I want to sit down and crank out a painting in the least amount of time, frame it and hang it on the wall. Reality is much different.

I began to notice the further into a painting I got, the more I began to freak out—almost to the point of not wanting to finish. In that panicked moment I realized if it didn't turn out the way I wanted it to, I'd have to start all over again. The more time I invested, the more time I would potentially waste. How could I justify this use of time?

While staring at the paper, not wanting to finish, I had a revelation. This tension I am feeling is very similar to the tension between production and programming (or any other ministry, for that matter). On so many levels, production just wants the information so we can execute it. From the creative side, they have to work through a creative process so an idea can become fully realized.

With watercolors, I have come to grips with that fact that I can paint the "same" picture four or five times before I figure out what works and what doesn't. Even after that, I could probably work and rework it until I am blue in the face. As an artist, I have to deal with this reality. I will probably never be fully satisfied, and it will rarely be an efficient use of time.

It isn't easy, but I can deal with this tension in my painting process because I only have to worry about myself. In the realm of the technical and creative arts, we have to work out this tension with each other. Creativity takes time. To execute an idea takes time. Neither one is more important than the other, and both are required to accomplish the goal.

In my early years of ministry, I would usually deal with this tension by giving in to whatever the creative people came up with, regardless of how it affected my world. Much of this came from feeling like I was there to serve artists and that I wasn't one myself. The other part came from my assumption that people knew what they were asking for. I wasn't working through this tension; I was ignoring it. My next step was to set up rigid boundaries that I wouldn't let break, no matter the circumstances. This wasn't working out the tension either. Yep, still ignoring the tension.

It seemed like a stalemate. How can this be solved? It can't. We can only work on it. It involves relationship. It involves communication. It involves not only understanding your unique contribution to the body of Christ, but your teammate's unique contribution as well.

We are called to work very closely together with this entire group of people we are potentially at odds with right from the start. However, when you get down to the bottom of it, we all want the same thing. For production, we want to be involved in something cool and creative and to do a great job executing. In programming, they want to create something next level and creative, and they want us to do a great job executing.

In order to get to this kind of outcome, we need to figure out how to understand each other's process. Just like we have been saying—nobody understands your world unless you tell them—we tend to misunderstand creative people's worlds unless they tell us. Lots of people struggle with the same issue of assuming that people know their worlds. Maybe for you to understand someone else's world, you need to ask them.

Here we are, called to work side-by-side with people we know very little about. In all the years of feeling like a victim, feeling like no one understood me, there were people on the other side of the table who were often feeling the exact same way. All this time I was waiting for someone to understand me and my world. It hadn't even occurred to me that I could be the one to reach out. Not only could I help others understand my world, but I could understand what the people around me were dealing with.

There is usually an unfortunate gap that separates us from the people God has called us to work with the closest. This is natural because God made each of us unique and different. But this gap is widened by misunderstandings, miscommunications and assumptions.

Chapter 24 Discussion Questions:

1. Read 1 Corinthians 12. What is your unique place in the body of Christ?

2. Think of a situation with another person where you came at the same issue from different perspectives. How can this difference benefit your church?

3. What are some differences between who you are and the people you serve with that you can celebrate?

4. What are steps you can take to improve your relationship with the people you serve with?

5. Name specific people and individual conversations that you need to have as a result of reading this section.

6. What are some past hurts where you need to experience healing?

25

BUILDING TRUST

If you are in a place where collaboration isn't happening, and you think trust might be the problem, what can you do about it? How can you build trust into the culture?

Since you can't make someone trust you by saying "Trust me," the only way to build trust is through your actions over time. How you respond in certain situations, how you treat people and the choices you make will either build trust or tear it down. I think most of us want to work in a trusting environment, so let's take a look at ways you can proactively build trust with co-workers.

Nail the Basics

As we've discussed previously, this is the foundation of production. But we can get so caught up in the new and shiny that we forget to do the basics well. If you are having trouble with these simple, basic ideas, then you aren't someone others want to collaborate with.

If simple and basic is difficult for you and your team, why would someone enter into a conversation about the new and exciting? If your worship pastor has concerns about whether a mic will be on or not, do you think she'd trust that some new exciting mic'ing technique would work?

As unglamorous as it is to do line checks, go through the lighting cues one more time to make sure there isn't a distracting moment, or pay enough attention to the graphics so they are coming up at the right time, these must be done for the foundation of production to be solid. Without these very simple and sometimes boring tasks, the rest of production isn't possible.

How committed are you to doing your best with the basics? The level of trust you feel from others will be a direct correlation to your ability to do these well.

Stop Blaming the Gear

There is no question that many times, better gear can yield better results. There is also no question that if you are always complaining about the gear you don't have or that the gear you do have is causing the problem, people's trust for you will be on shaky ground.

The best way to build trust with the people you need to collaborate with, whether the senior pastor or the worship leader, is to use what you have to the fullest. Stop complaining and get the most out of the gear your already own. This is straight from the earlier chapter about using what you have.

Get the most out of it. When there are problems, own up to them if they are yours to own. If gear failure in your services is a result of you struggling with the basics, don't point the finger at the gear. If your gear is causing the problem, make sure you talk about it up front. Let someone know you have concerns. Don't wait for it to fail before you throw it under the bus.

I recently volunteered to provide production support for a school event for one of my kids. It doesn't matter where I'm doing production or for whom, I want to do great work. I don't want people thinking about production, but about the content of the program. I want production to be as transparent as possible.

There were a couple problems right off the bat. The gear at this school is suspect at best. I never knew which mics would work or which lights turn on with which switches. Going in, I knew I was in for a challenge. The second issue was that the people presenting at the event wanted to use wireless headsets even though they were just standing at a podium the whole time. Knowing what I did about the system, I spent quite a bit of time trying to talk people into using a wired, handheld mic, since it would give us the highest chance of success. No luck.

You can probably imagine where this is going. Mic after mic failed. Either because of RF interference, a bad connector or the flimsy headset mic falling off someone's head. Fortunately, I had set up a wired mic by the podium just in case.

This example speaks to a couple of things. I knew that the gear was questionable at best, so I spoke up. Not to complain, but to explain what could happen. After the decision maker respectfully decided not to take my advice, we used the wireless mics. When they failed, I didn't make a big deal out of it, and I definitely didn't say "I told you this would happen!" I prepared for the chance that they might not work, and then I didn't carry around the weight of responsibility about whether it worked or not. I'd spoken up and someone else made the decision. I'm still going to do my absolute best to make it work, but if it doesn't, someone else made the decision and carries some of the responsibility for the distraction it caused.

I prepared for the chance that they might not work, and then I didn't carry around the weight of responsibility about whether it worked or not.

Building Trust Out of Failure

I've lost count of how many times I've recommended a solution to a problem, only to have the people in charge want to do the exact opposite. This used to drive me crazy! Trust is obviously very low if people are not only *not* taking my recommendation but doing the opposite.

To be trusted, I need to trust others.

I learned over time that these moments are even more defining for building trust than when someone takes my recommendation. If I predict that something won't work properly, but we do it anyway, this is another moment when someone can say "Todd told us this would happen, and it did. Maybe we should take his recommendation next time."

Now if I had jumped in with an "I told you so," I've blown my opportunity to build up trust.

By offering my opinion, then doing my best with whatever choice the decision maker selects, I've become a team player trying to accomplish the collective goal. I can be trusted to state my perspective, then do my best for the group regardless of my personal viewpoint. To be trusted, I need to trust others.

Next time, when I offer my opinion, people will remember last time and perhaps follow my recommendation.

How Trusting Am I?

The other side of being trusted is trusting someone else. To fully realize collaboration, I need to be willing to follow the leader's decision. I need to be able to release my ideas and my knowledge and my control issues to the group in order for us to make the best decision. This is not easy.

If I offer my opinion in a given situation, I hope my advice is followed, but if it isn't, I need to follow my leader. There needs to be no question in the leader's mind that if they are asking me to get it done, I will. For my leader to trust me, they need to know that I trust them also. For them to know that, I need to give them opportunities to be trustworthy.

Am I willing to put myself and my team in the hands of someone else? Will I make myself and my team available, or will I hoard those resources out of mistrust?

Trust Erosion

Sometimes trust has already been eroded. Maybe you've been too cavalier with trust, saying one thing then doing another, and no one believes you anymore.

Or maybe you've trusted someone and been let down again and again. What do you do in these situations?

It's probably time to move on.

I once worked with someone who had broken trust so much that there was no way to build it back up again. So much trust had been lost that no one would have believed long-term change was possible. Ministry moves pretty fast and sometimes the bridge is burned, so instead of trying to rebuild it, it's time for a new bridge.

Maybe this has happened with someone you lead. It's time to let them go. Whether they are a staff member or a volunteer, you can't continue to drag the rest of your team down by having a person on the team who can't be trusted.

What if you work for someone who doesn't seem to trust you anymore? First, have a conversation. Find out if there is truly a lack of trust or if it's just your perception. Maybe it's a misunderstanding. Ask how you can begin to rebuild trust. In most cases, the rebuilding process begins by doing what you say, stating your opinion in helpful ways, then executing whatever the plan is to the best of your ability.

On the other hand, if there is no chance to rebuild trust, it is time for you to choose to move on. If you don't, chances are that you will only continue to be frustrated and stymied at every turn.

Let's face it, when you're not happy, you aren't the best version of yourself anyway. Take the huge leap and step aside. You're missing out on what God might have for you because you're too afraid to let go of something you are probably holding onto too tightly.

Whether you are building trust from the ground up or you're trying to rebuild trust, this is not an overnight experience, and it isn't easy. Begin the process now.

Chapter 25 Discussion Questions:

1. I know, I know ... the basics! Can you be trusted to do everything possible to nail the basics?

2. When problems happen, are you quick to point the finger at everything else? What piece of gear do you most often blame?

3. Think of the person you tend not to trust. Why is that? What baby step could you take to show your trust in that person?

4. Can you identify a point in time when trust started to erode? Who should you have a conversation with? What is a concrete step you can take toward rebuilding the trust?

26

CONTENT IS KING

I want to make something as technically excellent as possible, and I don't necessarily care if it is actually what our service needs. If it is the wrong song for the moment, but it sounds and looks great, I tend to feel like the service went great.

Writing it down for this book, it is kind of embarrassing. Really, Elliott? Lame. However lame it might seem, I think that is probably where most of us tech people live. If it isn't exactly accurate for you, there is a good chance that the creative people you work with believe it is true of you.

But the reality is that none of the production matters if we are supporting the wrong content. In the context of any live production where tickets are being sold, nobody will buy tickets to a Journey concert if the mix sounds amazing, but they only do new songs that nobody knows. People are coming to see Journey for the content, not the amazing light show. If I don't hear "Don't Stop Believin'," who cares if the IMAG is in SD or HD; I'm going to give my experience a bad rating on Stubhub.

Once we had a guest speaker at church whose flight was delayed, and he missed our Saturday night service. At the time, all of our regional campuses were taking the recording of that service and replaying it the next morning.

As we started to brainstorm what to do about it, one of my very bad ideas was to have the campuses use our beta testing web stream on Sunday morning. (This was a long time ago!) The immediate reaction from the TDs in the room felt like outrage. The stream looked bad and couldn't be blown up on a screen. Our bandwidth wasn't enough to handle the quality of image that is acceptable. (Again, this was very long ago!)

I would have to say that I agreed with each point they were bringing up. But none of them got at the heart of what I was feeling. If it was important for the congregation to hear the content, the method of delivery was secondary.

This might seem like I'm contradicting my earlier stated value of doing our jobs with excellence, but it isn't. If the content is valuable, what do we have at our disposal to deliver it? Given the circumstances, how can we do this with excellence?

At the end of the day, we ended up going with a different option, but my idea was more about the importance of the message than the quality of the delivery. Having everything be technically excellent is an important value, just not at the expense of the quality of the content.

If there is a message on an old wax cylinder record that your pastor decides the church must hear, what is the best way for us to play it? If there is a video that explains the church's strategy for the next initiative, but it only exists on a VHS tape, how can we make it the best it can be?

If it was important for the congregation to hear the content, the method of delivery was secondary.

This is where the idea of trust comes in. If there is trust that you will do your best to make the content happen, and if there is trust that the leadership understands the limitations, but feels like the content is strong enough, then this is another opportunity for collaboration.

There was a season when it seemed like the production team's primary goal was to have a great rehearsal. If everything went according to plan and we rehearsed everything thoroughly so that we could execute the plan with excellence, then we were happy.

In reality, a smooth rehearsal shouldn't be the goal. The trouble with creativity is that you can't always figure stuff out on paper. Sometimes you can't know until you are seeing and hearing things for the first time. Other times, based on a subtle, last-minute change in the pastor's message, a different song would more accurately capture the main idea.

While a great rehearsal is a good idea, it isn't the most important idea, especially if you are rehearsing the wrong content. There were more than a few times when we would throw out the entire worship set after the Saturday night service. So much for an amazing rehearsal.

This is a dance we need to learn as technical artists. We need to push for content as soon as possible so we can plan. For the sake of our volunteers and for our congregation, we need to drive toward a well-ordered rehearsal so we can create an environment where people can meet with God. We need to do everything we possibly can ... and then we need to be ready to let the content take priority.

Chapter 26 Discussion Questions:

1. How can you change your mindset on content over quality?

2. Is there something in your current process that is choking out good content?

27

IF YOU CAN DO IT, DO IT. IF YOU CAN'T, DON'T

I don't know about you, but the closer rehearsal gets, the busier I tend to get. Cleaning up messes, correcting typos, doing that thing I forgot to do on Thursday.

Typically, these are things on my list—things that I want to get done because I generally don't do things that are on other people's lists. What's not on my list is the band leader who shows up with an extra guitar, an extra vocalist and a song change. When that happens, it can be easy to blow a gasket.

How am I going to be ready for rehearsal if I don't get my stuff done? "Don't you know that I have enough to do without you adding to my already big list?" When I find myself in this place, it is easy to freak out and just say "No! I can't do these things."

You can't say it's impossible and then turn around and make it happen.

Many times I've said no immediately and then after thinking about it for a while, I have found a way to make it happen. I would then do it even after I said it couldn't be done. I can remember one time when I said we couldn't add an extra tin whistle, and then I ended up providing a mic for it. My boss called me on the carpet for it. "You can't say it's impossible and then turn around and make it happen."

Trust—Down And to the Left

If you say one thing and then do another often enough, no one is going to believe you.

Nothing drives trust out the door faster than saying something can't be done, then proving the exact opposite. Trust is the key commodity for true collaboration to happen, and if you are doing things begrudgingly that you've already said no to, you will lose trust. Once it is lost, it is very difficult to get back.

From my perspective, I'm usually pretty proud of myself for figuring out how to do the impossible, and I usually want others to be amazed by what I've pulled off. Instead, the exact opposite is happening. Slowly over time, your ability to do the impossible after you've already said it can't be done erodes trust.

Give Yourself Time to Think

Instead of responding immediately with the negative, ask for some time to think about it. In the pressure of the live event, this can be pretty difficult, but by not answering right away and asking for a minute, you've communicated what you really need is time to come up with options.

Sometimes there isn't time, but don't let that stop you from asking. That way the person asking for the impossible knows that you don't have a solution right this second, but you might be able to come up with one if you're given some space.

Let's say there isn't time to wait, and you've said you don't know. Then the answer needs to be no! The situation is to blame not you. This doesn't erode trust at all, but instead, builds it.

In this example, what if you then came up with a solution after some time had passed? Do you think the other person will be angry that you came up with a solution after the fact? No way! They are impressed by your brilliance! How did you come up with a solution to the impossible?

By asking for time, you are communicating. Communication is key to real collaboration with creative artists.

If it Can't Be Done, Say No

I like to be a team player. I don't like to let people down. Many times, something can't be done simply because you or your team don't have any more time ... at least not a normal amount of time.

I've noticed that it is easy for me to sacrifice my time for an idea. I don't always consider my time to be a resource. Because I don't want to let my teammates down, I say yes to something instead of having dinner with my family or going to my daughter's dance recital or finishing another project that isn't as urgent, but is more important in the long term.

You are not helping your team any by not speaking up for the things that you don't have capacity for.

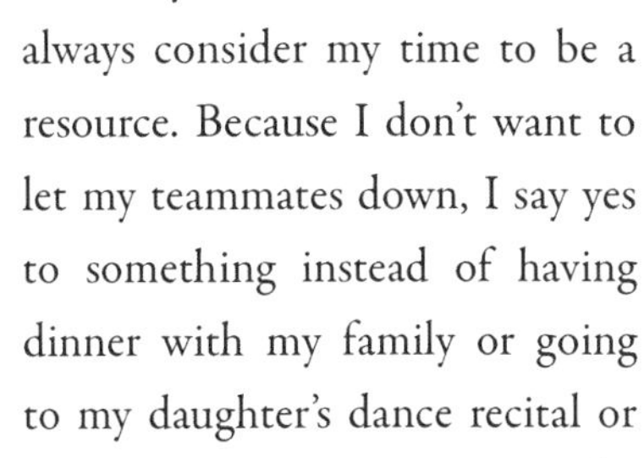

Don't be afraid to say no for reasons that seem squishy and selfish. Your time is valuable, and it matters. Your team needs you to speak up. If you don't, pretty soon you are going to be bitter for all the ideas that "force" you to miss family time. Remember the grumpy tech person? It might be you. Don't let this happen in the name of being a team player. You are not helping your team by not speaking up for the things that you don't have capacity for.

If it Can Be Done, Say Yes

Many of us are great at *trying* to set boundaries around deadlines and how soon we need to have information. Most of this comes from being burned at the last minute. The problem is that once you put a deadline out there, you feel like you need to stick to it when people don't meet it. "You didn't get me the graphics two weeks in advance, so I can't do them for you."

Deadlines are set up as a framework for the tasks we need to get done. The deadlines don't drive what we are about, they serve it.

If you have the capacity, why not help get it done? Deadlines are set up as a framework for the tasks we need to get done. The deadlines don't drive what we are about, they serve it. If my pastor comes to me at the last minute with something that he already knows has a deadline associated with it, why wouldn't I try to make that happen? There should already be an understanding that since the deadline is passed, you might not be able to get it done. To make sure there is that understanding, consider mentioning it.

Fight the urge to fold your arms and refuse to do something you have capacity for just because someone missed a deadline. If the service will be better with the addition, and you can do it, why wouldn't you?

Chapter 27 Discussion Questions:

1. Do deadlines drive you or are they more of a guide?

2. Do you jump to the idea that something is impossible before considering the options? Develop a statement that gives you more time to think.

3. What boundaries have you set up to protect family time?

28

NOT ALL IDEAS NEED TO BE FIGURED OUT

When I was invited to be part of the brainstorming meetings for our services, it wasn't long before I was kicked out.

As a production-minded person, I would immediately go into problem-solving mode with every idea. Since I was the one who would have to make the ideas happen, I tended to not like most of them. I would start shooting holes into everything. And they were pretty legitimate holes. Not enough gear. Not enough time. Not enough money. Not enough whatever.

Since those days, I've noticed something about brainstorming. The majority of ideas are just that ... ideas. Right at the beginning of brainstorming, the chance that these ideas will actually be the thing we do is pretty slim.

For anyone who has written anything, you know that the most important thing you can do is just get something down on paper. Don't overthink it, don't try to solve every problem while you are trying to formulate the ideas, just write it. Write it badly. You can always go back and fix it. And more than likely, it will change drastically and not look the same as when you started.

Create a Safe Place

This same idea needs to be in place during service planning. For there to be a good service plan at the end of the meeting, the team needs to go through a lot of bad ideas before finding the one that will work. If I shoot down every bad idea that gets thrown on the table, I'm not creating a very safe place for people to share their ideas.

Ideas are fragile. The people sharing those ideas are also fragile. Remember how much you like getting critiqued? It is important to learn when it is time for solving problems and when it is time to just relax in the sharing of crazy ideas. Like I said earlier, the chance that any one of those ideas will actually happen is pretty small.

Production is a Part of the Brainstorming Process

Let's say the group has landed on an idea. I've noticed that many times the brainstorming ends, the marching orders are given, and we all leave the meeting. In reality, figuring out how something will actually be accomplished is the necessary next step in the brainstorming process. If an idea can't be pulled off, then it is still just an idea. As the production representative, it is your job to help figure out ways to turn the idea into a thing.

This is where your expertise really kicks in. You are the one with ideas on how to make the idea a reality. That is what you have been created for. Feel confident that this is your role and that it is necessary. Don't hold it over people's heads, but don't shrink away from the fact that this thing needs to get figured out.

Constraints help to sharpen thinking and even make an idea better. Regardless of your perception of the resources other churches have, there are always limits to what can be done based on the money, time and people available. The challenge with production brainstorming is pulling off the essence of the idea with the resources you do have.

I used to feel like I was destroying people's ideas with my lack of resources. In reality, what I have is what I have. It is my job to figure out how to fit the idea into what can be done.

We've talked about trust between the technical and creative artists and what a vital commodity it is. Well, this is where it pays off. To enter into a production brainstorming session with high levels of trust starts things off on the right foot. When trust has been built, it turns the conversation about limits into one about figuring something out together.

Turning the Idea into a Thing

If we didn't have the resources to do someone's exact idea, I usually just told them it couldn't be done, and then I expected them to come up with a new idea. This was before I realized how difficult it is to have an idea in the first place.

When I approached brainstorming this way, I was not a team player. In fact, I was quite the buzz kill. I wasn't providing solutions—only pointing out problems. The amount of time and effort that it takes to come up with a new idea is not something to take lightly. If something can't work with what we have, how can I help shape the idea to fit what we can do?

If it's not clear to you already, nobody knows the world of production like you do. What better person to help figure out how to make an idea work than you?

When you first hear an idea, and how production plays into it, you are only skimming the surface. It is based on what someone else imagines can be done with production. When you start imagining how to accomplish the idea, you are drilling down to the essence of the idea. In my opinion, any time you can simplify the idea into its primary components, you are stripping away the unnecessary parts and getting to the heart of the idea.

It's in the heart of the idea where you are able to see more clearly what production can add or subtract to bring out the essential part of the idea. Just writing this down I get excited! This is where the fusion of the technical and creative arts really come together! This is what God had in mind when he created you and put you on the team you're on!

Limits are Good

I believe that ideas will always push the boundaries of what you have available. Let's pretend you had all the resources in the world. Production without

Ideas will always push the boundaries of what you have available.

resource constraints is typically overdone. With unlimited funds, things tend to go over the top quickly. Limits are frustrating, but they make any production, any service, any idea better by the hard work of thinking through each aspect of an idea to determine the essentials and eliminate the fluff.

When it is time for brainstorming, relax. Confining brainstorming to just the creative idea isn't enough. We need to expand our definition of brainstorming to include how we are actually going to execute the idea.

Ideas are great, but unless they can be brought to life, they will still just be ideas. Lean into what you know about production to help get your team's creativity off the ground. Your team needs you to relax and create.

Chapter 28 Discussion Questions:

1. Does unlimited brainstorming stress you out? What are some ways you can exist in the unknown and not have to figure everything out in the moment?

2. Connect with your creative arts team about how to make their ideas happen. Develop ways to communicate limits with each other.

29

GET SOMEONE ELSE TO SAY NO

Unfortunately, the tech person at the local church is well known for saying No to crazy ideas. Being the No person is not a fun place to be. I used to think that I was being a realist. I was the only one who could see how crazy the ideas were. For those on the other side of my no, I was basically impossible to work with. (Sorry. You know who you are.)

The reality is that the creative people you work with have no idea how to accomplish their ideas technically. That's what you are there for.

When it dawned on me that I was there to help *realize* creative ideas, and they were there to *create* ideas, I started to think about the word "no" a little differently. Instead of rejecting someone's idea right away, what if I could get them to say no for me?

Some of you are thinking that this is where we turn on the classic passive-aggressive tech person routine. Wrong. The exercise of trying to get someone to say no to their own ideas was a creative way of problem solving together.

Show Your Work

Remember when you were in math class, and your teacher would take points off your test if you didn't show your work? You couldn't just write down the answer; he wanted to see your process and how you got there. It was a way for him to know that you understood the material.

When we lead with the answer No, we are not showing our work. For people who don't generally understand the world of production, you are keeping them in the dark by not explaining how you got to No. Part of what keeps us from sharing all the details is that we assume one of two extremes: that most of it is over people's heads, or they already know how impossible their idea is and are asking for it anyway. My guess is the truth is somewhere in between.

When I am working with my kids on homework and pushing them to show their work, they usually get defensive about how much they know about the subject and that they shouldn't have to show how they got the answer. But without the details, I'm not totally sure they understand the concepts.

How often am I defensive about showing my work? What if they think I don't know what I'm talking about? What if I don't know what I'm talking about? When you open yourself up to someone critiquing your thought processes, you are also giving them a glimpse into what is involved in executing their idea.

Provide Options

As you are showing your work, give your creatives a chance to find a solution with you. Unlike your middle school math problems, most creative ideas have more than one answer. Also, unlike your middle school math problems where you are showing your work to demonstrate that you know how to get to the answer, showing your work in a production sense is more about discovering the answer together.

In my earlier years, I thought it was my job to come up with the answers myself. It was my responsibility to have the answers. Looking back, when I would come at the problem this way, the answer would usually be no because, based on my understanding of the idea, I couldn't see how we could get it done.

Eventually I realized that pulling off an idea wasn't all my responsibility. It needed to be a team effort, and me saying no all the time wasn't fostering a team experience.

I figured out that it was my responsibility to provide options for how we could get this done. Giving three or four possible ways of accomplishing an idea helped share the weight of figuring out how to make an idea happen.

By not saying no immediately and providing a few ideas, you've now become part of shaping the idea. You are helping to refine it into its best version. If you can't execute on an idea, it will never be anything more than an idea. Using your expertise in production, you can share the burden of making the idea happen.

Showing your work in a production sense is more about discovering the answer together.

When you are providing options, you are asking for the idea person to help choose which solution will accomplish the idea the best. We can do this, but we need extra time. We can do this, but it will cost something in rental. If we had two more qualified volunteers, we could do it. If we altered the idea slightly, we could use the resources we have and pull it off.

These kinds of options give people a glimpse into what it takes to do live production, and it also shifts some of the burden of figuring out their idea onto them.

When providing several options, it's important that the options you talk about should actually be doable. You shouldn't offer up a solution that you wouldn't actually recommend. For instance, if the solution is needing time to replace a piece of equipment, don't offer another option that's so cheap it doesn't have the features you need to do the job at hand. By providing good options, you are now a team player trying to make someone's idea a reality.

Share the Decision-Making Responsibility

Sometimes an option involves choosing between two important ideas. Or sometimes between two bad options. Should we do this or do that? It's important to get input from others on these kinds of options.

I've been reading the book "Creativity, Inc." by Ed Catmull of Pixar. They assign a certain number of "person-weeks" to a project: the amount of work that one person can get done in a week. Each one of the person-weeks is represented by a popsicle stick, and each project only has so many popsicle sticks to work with. If the director wants more time spent on a particular part of the project, she needs to make the tough decision of moving the popsicle sticks from one part of the project to another. They don't have unlimited person-weeks at their disposal, so it's the leader's responsibility to decide.

In my earlier years, I tended to assume I knew these answers. The reality is that there are only so many ideas that can be accomplished with our limitations. There is only so much money, and there is only so much time. Providing options and brainstorming the production solutions helps share the burden of responsibility with the creative or the leader who is asking for the element.

If they want it to get it done, and they can come up with the money, let's do it! If they want their idea made into reality and can move other priorities around to give us the space to make it happen, I'm in.

Without a conversation about what's most important, what's the best way to accomplish something, or without someone needing to make a value call, the ideas still might happen but at the expense of people. And without people, we're back to ideas that won't happen.

I've had the privilege of leading a production team on a few trips to Germany to produce a Leadership Conference in a rented arena. It is tons of work, but also some of the most meaningful ministry I've been a part of.

At one of these events in the early 2000s, on the morning before the first session, I got an email asking for video footage from a baptism service a couple of weeks earlier. I started asking around to see if it was possible, and people quickly started freaking out at the technical impossibilities. Before we pulled the trigger to make it happen, I wanted to gather all the information and let the leader decide what we should do. I have to admit that my first thought was, "You do realize we're in Germany?" But I kept that to myself. Remember now, early 2000s. The internet was not what it has become. This was going to be a huge undertaking given our resources. This was before the iPhone, people!

I went back and presented the options of waking someone in Chicago to get the footage (because of the seven-hour time difference) and sending it to us over the internet, which would probably take too long, or maybe we could get some photos of the baptism. Once the leader heard what all the options were, he said, "Oh. Let's not do that; it was just an idea."

Don't take responsibility to figure out which ideas should be done, and don't take responsibility to make every idea happen the way it is first presented to you.

By providing options, we were helping him understand what he was really asking and how we might get it done. We were ready to do it, but by letting the leader in on what was involved, he had a chance to weigh the importance of the idea. If we are going to wake someone up on the other side of the world to do a task, the leader needs to be the one making that decision because it affects the organization's perception of him. If I'm making that decision for him, I'm contributing to a perception of the leader that might not be true.

Don't take responsibility to figure out which ideas should be done, and don't take responsibility to make every idea happen the way it is first presented to you. It is your job to tease out the necessary details to make the idea doable. It is not your job to carry all the responsibility to make the idea happen. Share the responsibility with your creatives and your pastor. They are the ones who need to make the final call about what should or shouldn't be done.

Chapter 29 Discussion Questions:

1. Are you known for always shooting down ideas? How can you begin to turn that perception around?

2. Do you show your work? How could you start including people in your thought process before you just say no?

3. How good are you at providing options? Think of a current scenario and develop three quick, realistic options to present to your team.

30

CHANGE HAPPENS

By now, you're dialed into the fact that I love to paint ... not my house, but with watercolors on paper. Nothing has given me more insight into the process for creating services than living through my own personal creative process.

Without fail, whatever I imagined for a painting never happens. Along the way, I have to make adjustments as my ideas weren't executed like I wanted or something unforeseen happens. These changes come in two forms; one I have control over and one that I don't. In either case, at the end of the day, I am trying to make adjustments so that the painting is the best it can be.

In the world of production, this was an eye-opener for me. In the planning phases of any service, I wanted to nail down every detail so I could come up with a plan. There weren't many things that frustrated me more than when those plans changed. I used to think people were changing things for no good reason ... at least not a good enough reason.

Things We've Planned to Do

Back to the painting analogy, when you work with watercolors you need to plan what you want the painting to be, because you can't paint over a mistake or add things later. You need to know where you're going and head in that direction.

My ability to plan out a painting correlates directly to my experience. How many paintings have I done? This actually means how many mistakes have

I made in the past? When you translate this to creating services, the same applies. We need to have a plan for what we want our services to look like, and our ability to get this right is based on experience—how many mistakes we've made in the past and what we've learned from them.

For you and your team, it is important to know you've prepared your best and that things might still need to change.

When we enter rehearsal with our plan, what do we do when it isn't working? This is where it is time to consider changing the plan. At this point, it doesn't matter how well we've planned, or how much we think the idea should work. If it isn't working, it is time to let go of the plan emotionally and try to fix what isn't working.

This can be very difficult on the production side, because it can mean that all the work you've invested is for nothing. Maybe more importantly, your volunteers have killed themselves to pull off Plan A, and how are you supposed to tell them to undo it all and start over?

It helps to start from a place of assuming that things may need to change. It also helps to develop the best plan possible. We don't want our lack of planning be the reason for last-minute changes. For you and your team, it is important to know you've prepared your best and that things might still need to change.

When you are executing an idea that has never been done before, there is always a chance that things will need to change. Have you planned the best you can? Are you ready to make last-minute adjustments so the service is the best it can possibly be?

Sometimes there isn't time to make a significant change, or the risk of changing something at the last minute isn't worth it. This will require common sense on your part. In the world of watercolor, it might be a change for the next painting, and I put it in the category of "learning."

Things We Haven't Planned to Do

In the creative process, there are always things that come up that nobody thought of. A light going out at the wrong moment. The audience responding in an unexpected way.

In the world of watercolor, the water has a mind of its own. If you aren't willing to just go with it, you should probably start painting with something you have more control over.

This kind of change can be thrilling to be a part of. It is about responding while something isn't going according to plan. It is about being alert to what's happening in the room, to get your head out of the cue sheet and be fully in the moment.

As a production person, especially one in the local church where we aren't doing the exact same show night after night, changing in the moment is the name of the game. If you aren't into it, you should probably get out of it. Many of the most memorable moments I've experienced have been a result of these unplanned and unforeseen events. Will I be open to them happening, or are we going to stick with the plan?

We once had a service where the worship time was blowing the roof off the place. In the moment, someone from the congregation came up on stage and wanted to sing. The team scrambled like crazy to get her a mic and make adjustments, trying to make it happen. When she opened her mouth to sing, we were all stunned. It is a moment I remember so vividly, because in that unplanned moment, the worship rose to an entirely new level—one that would have never happened had we not been willing to go "where the water was flowing."

Being able to adjust in the moment is a skill that has to be learned. It requires mastery of our craft, and our ability to work together on the fly—to understand what needs to happen and then doing it together.

Change will happen. Whether it is because what we plan isn't working, or because something needs to change in the moment, we need to accept that change is coming. How will we respond?

Chapter 30 Discussion Questions:

1. Is your team known for making changes because of poor planning? How can you adjust your process to help minimize this kind of change?

2. How willing are you to make changes to make the service better, even though it isn't what was planned?

3. Are there ways that you could prepare your team to make changes in the moment?

31

BECOME A LINGUIST

I'm a US citizen, which means a few things, but for this example it means that I only speak English. I know one language. I haven't taken the time to learn someone else's language, because fortunately for me, much of the world speaks English.

I'm not proud of my laziness. I can understand some German. "Ein bisschen." But for the most part, I am privileged to speak in the language that I understand best. For many of us in the technical arts, the language we understand most revolves around production terms and model numbers: SM58. H.264. DMX to Artnet Converter.

If collaboration is the name of the game in what we do, I need to learn to communicate in a new way that doesn't involve my most comfortable language.

You and I might understand what these terms mean, but many of the people we collaborate with have no clue what any of it means. Since I'm most comfortable talking in my native language, I do. And it doesn't take long to lose my audience.

If collaboration is the name of the game in what we do, I need to learn to communicate in a new way that doesn't involve my most comfortable language.

Translate

As we work with creatives, we need to take all the information we know, in the language we best understand, and we need to convert it into something that makes sense to our audience. High level theory about color temperature or talking about degrees Kelvin doesn't help the conversation. These ideas need to be put into a context that a non-technical person can understand.

Do we want the lighting to look warm or cold? This type of language makes more sense to someone who doesn't live and breathe lighting color theory and it gives them the context they need to help answer the question.

Your worship pastor probably doesn't care about whether you use an SM58 or an AT2010. She just wants vocal clarity in the mix next week.

It is our job to figure out what language the person we are collaborating with speaks, and then try our best to translate what we know into that language. If we want to be understood, it is our responsibility to figure this out. We can't assume that everyone "speaks production." For us to get the results we need, we need for others to understand what we're saying.

I find myself quoting the line from the movie "Rush Hour": "CAN YOU UNDERSTAND THE WORDS COMING OUT OF MY MOUTH?" When Chris Tucker's character thinks that Jackie Chan's character can't understand English, he just talks louder. If someone doesn't understand what you're saying, saying it louder or more often isn't going to help their comprehension.

For those of us communicating technical information to lots of different groups of people, it isn't enough to just speak one other language, we need to become multilingual: Worship-Leader Speak. Senior-Pastor Catchphrases. Children's-Ministry Enthusiastic-Phrasing. Youth-Pastor Lingo.

More Than Words

In order to collaborate and to accomplish all we can to help the mission of our church, it is our responsibility to learn how to translate what we know into something other people can understand.

This doesn't just apply to the words we use, but also to concepts. Our reasoning behind why we need to upgrade our projectors might not make sense to our pastor. Don't even get me started on the 600 mHz change over. Nobody, including me, understands that!

When talking with your pastor, know that he needs to understand the *why* behind what you're saying. We need to translate the needs of our ministry into how it will help advance the church's mission.

We aren't just replacing all our front light with LEDs because it's cool; we're doing it to help save money on power consumption, hence bringing our operating expenses down. That's the language of a senior pastor!

We need to replace our sound system, or we can increase vocal clarity so the spoken word is easier to understand in more places.

Whatever we are trying to communicate, it is important to connect it to an idea or a value that the church believes in deeply.

The Curse of Knowledge

You know how frustrated you get because nobody understands your world? That isn't going to change unless you can help educate your co-workers on what you do and how something can get done in terms they can understand.

It can become easy for me to assume everyone has the same knowledge base as me. The reality is much different. Nobody knows exactly what you know. We need to start communicating with people with the understanding that not everyone knows what you know.

We've been living with a particular idea for longer than anyone else. We've got years of previous experience that have brought us to a certain conclusion that, chances are, nobody else has come to. It is only natural that we're farther down the road on a concept than the person we're explaining it to. Use more words to help people understand.

This goes for the volunteers on your team who are being trained, as well as the many other non-technical people you are trying to work with who don't understand all that goes into what we do.

This idea of the curse of knowledge got me into trouble when I was first starting out. I made lots of assumptions that people knew how difficult the tasks were or that they knew I would have to stay until late into the night to get something done. I'm confident we've established the fact that nobody really knows what we do, and it is up to us to share our knowledge so that others can understand.

For us to move the mission of the church forward, it is key that we learn how to translate our ideas into words and concepts that the people we are collaborating with can understand.

Chapter 31 Discussion Questions:

1. Think about the people you collaborate with. How many different languages do you need to learn to speak?

2. What is most important to these people? How could you frame your next conversation into these concepts?

3. Think about the beginnings of your production journey. How would you explain concepts to your younger self?

32

RULES OF IMPROV

I love to be back in the booth. I prefer to not be in the spotlight. In fact, if you can create an absence of light, I wouldn't mind hanging out there. Knowing this about myself, it is no wonder that I really hate the game of charades. Not only is it unpredictable, but I have to stand up in front of people at the same time. This is not a good combination.

Combine charades with the improv game show "Whose Line is it Anyway?" I thought the show was super funny, but it mostly made me sweat. A lot. That's a ton of unpredictable, and it's on TV! No thank you!

That being said, we have quite a bit to learn from the world of improv. Most of us can tell when it isn't working quite right. All you have to do is watch that episode of "The Office" when they follow Michael Scott to his improv class, and his answer to every scene is to pull out a gun. Scene over.

In the world of production, we are faced with situations in which we need to respond in the moment; where we have to make a choice whether the "scene" continues or ends right there. In the local church, we are trying to pull off something relatively new every seven days, which means we are brainstorming about new ideas non-stop. In these environments, are we helping move things along or are we following Michael Scott's example and pulling out a gun?

Let's take a look at a couple of essential components in the practice of improv and apply them to our context.

Accept Every Offer

In improv, the idea is for each person to respond to what just happened. Not with a gun and not with a question that just puts the scene into the other person's court again. The gun ends the scene, and the question makes the other person do all the work.

Since God wired the technical artist to figure out how to get things done, we tend to go into solution mode the minute we hear a new idea. Many times, this is perceived as pulling out a gun or putting the burden back on the other person.

So how do we handle this? If we're designed as technical artists to operate a certain way, are we trying to become something we aren't? How are we supposed to figure out how to make something work if we can't start picking it apart and pointing out all the flaws?

Well, one place we can start is by responding to each new idea with our reaction to the idea itself. Communicate that you like the idea, recognize how much work the person put into it, or point out all the great components to the idea. These are all better places to start than, "That will never work."

I'm not suggesting you lie or make something up, but remember, from a creative standpoint, it is difficult to come up with new and imaginative ideas. Whether it seems this way or not, your creative people are putting themselves way out there every time they present a new idea. They are opening themselves up to being laughed at or ridiculed.

We need to be problem solvers, not just problem pointer-outers.

They are also opening themselves up to collaborating with you. In most of our churches, ideas can't happen without production playing a significant role. The creative team knows this, and that is why they feel compelled to share their ideas. Are you making it easy for them, or do they dread meeting with you?

At its heart, collaboration is a partnership, much like improv. Improv doesn't work if one person is always dominant or if one of the partners is floundering. And we can't just respond to ideas with warm fuzzy sayings. We need this thing to go somewhere; we need to dig into the details and figure out how to make the idea happen.

I've met so many church tech people who respond to every idea with a "no" or "it can't be done." These classic examples are a great way to stop collaboration from happening and shutting the whole thing down. It isn't bad when an idea isn't doable, but we need to show up with solutions. We need to be problem solvers, not just problem pointer-outers. Instead of the phrase "That will never work," what if we tried, "How can we make this work?" While this is a question, it helps us move the scene along.

Make Your Partner Look Good

Going back to the rules of improv, the second big idea is to always make your partner look good. As I said before, if only one person looks good, pretty soon the improv scene is going to collapse.

When we are talking about collaborating with creative artists, we need to figure out ways for that the idea to work. When we are in brainstorming mode, it is easy to start solving the problem within our current reality instead of opening up the possibilities and throwing out any option that will help the idea become real. It could be renting more gear or hiring people. It could be tweaking the idea slightly to fit inside the gear and people we have. But what if we thought outside our normal box?

If God designed the body of Christ for us to work together, and I believe he did, we are in this together. It might be an oversimplified way of saying it, but "making your partner look good" is about combining our gifts and talents with others to create the best thing possible. Not only should the starting point be one of trying our best to make someone's idea amazing, but it should also involve trusting that the person with the idea wants the same for us.

In the same way improv requires mutual trust, so does collaboration. If you don't trust each other, it is difficult to put yourself out there. The reality is that the creative arts team is designed to imagine new ideas, and we are made to help make them a reality. We are designed to work in partnership together. If we hope to create life-changing moments through the fusion of the creative and technical arts, we need to accept every offer and make our partners look good.

Chapter 32 Discussion Questions:

1. What is your initial response to a new idea? Is this helping or hurting collaboration?

2. Are you a problem solver or a problem pointer-outer?

3. Think of a situation where collaboration broke down. How could you have responded differently to create a different outcome?

PART 4

PRODUCTION LEADERSHIP

33

TENACITY TO LEAD

Everything we've talked about up to this point has been about you as an individual tech person: how you think about production and how you relate to other artists you work closely with. Most of what you need to know to thrive as a technical artist is contained in the previous pages.

Learning to be tenacious with the basics of production is the foundational work we all must do in order to bring production up to the level our churches deserve. Going after each relationship and working hard at collaboration are keys to the next step in our development toward being fully functional members of the body of Christ in our churches.

Some of you reading this are also leading a production ministry at your church. And if you are anything like me, you are also a tech person and you have been "hands-on" until this point. As your church has grown, you have needed to take on more of a leadership role. Whether it is leading staff or volunteers, figuring out how to go from doing the task "hands-on" to mobilizing others to get the work done can be a huge leap.

When I first started making the transition from operator to leader, I felt like I wasn't doing anything. I had spent many of my early years working in one of the production disciplines, getting it going, then handing it off to the next person. Once the hand-off was complete, I would move onto the next discipline. At a certain point, I ran out of disciplines, and my boss told me I could not pick up some new task. Instead, I needed to focus my energy on leading the team.

Being a leader of tech people, frankly being any kind of leader, requires a great deal of tenacity: a very different type of tenacity than you've had to deal with so far. Being tenacious with the basics of production and in relationships are all about things that you have direct control over, whether it is your response or how you're going to handle yourself in a given situation.

When you start talking about leading a team of technical artists, you are now transferring what you know and how you handle each situation to a group of people—individuals, with minds of their own. They have opinions on how things should or shouldn't be done. This group can choose to follow your lead or not.

Making the transition to becoming a leader can be one of the most difficult challenges you will experience. It involves losing a little bit of your identity as a technical artist. If you thought you were in a "behind-the-scenes" role before, now you are a leader who is even more behind the scenes, working toward giving your team what they need to succeed as technical artists.

Of all the things requiring tenacity, this is the one that I think is the most difficult.

After one Christmas program at Willow Creek, we were making many of the resources from the service available to other churches so they could reproduce the service.

As part of that effort, we were putting commentary from those involved in the Christmas production on the DVD. (Remember those days? DVDs!) There was one commentary from the music people and one from the lighting designer, another from the drama team and a fourth from the set designers. I got an email from the people putting together the commentary groups asking me to give my own version of events from the aspect of a production.

As they started the video of the service, I realized that everything I wanted to say had already been said by one of the other groups. So, I started thinking about how I had cleared the way so people could fully contribute their uniqueness to getting the event done. But you can only say this so many times ... and nobody really cares about that part of the behind-the-scenes.

Needless to say, they didn't use my commentary.

Not only do you need to be relentless with casting a vision and creating a sense of team for your staff and volunteers, but you actually need to accomplish something. And while you're motivating your whole team toward a goal, you need to remember what you believe and why you believe it so you can keep moving the ball down the field.

Leadership can be a lonely place, as many of you have discovered and many more will soon find out. I often have the familiar statement "first in, last out" in my mind as it pertains to technical artists. For the leader of this group, it can be even more extreme. You are the first and the last of the "first-in, last-out" group.

In this lonely place, you need to learn how to lead yourself to keep moving forward. If you can't keep going, you can't expect to be able to lead your team forward.

Leadership will be the most difficult thing you'll do, but it can also be the most rewarding. When I think back on the highlights of my ministry so far, they all involved watching the teams I've led do amazing things together.

At a certain point in my own journey, I began waiting for someone else to pick up the leadership baton. But then I figured out that everyone on my team was looking at me. I love what I've been able to do and seeing how God has used my experiences to help my church grow and become better along the way.

As someone who has gone on the journey from operator/booth-sitter to leader of tech people, I hope that this section on leadership helps you make the transition. You need to do this not only for your own sake, but also for the sake of the teams that you lead. It takes guts, and hopefully the ideas that follow will help give you a practical road map to becoming the leader your team and your church need you to be.

Chapter 33 Discussion Questions:

1. Has it been difficult to let go of "doing" production? What are some things you're still hanging onto?

2. What is your biggest growth edge related to leading your team?

3. When you think of leading your team, what gets you the most excited?

34

MISSION, VISION, VALUES

Now that you're leading your production ministry, whether you have an all-volunteer crew, a small staff or a large one, it is time to get this group organized. And it is not just setting up systems for them to accomplish the task well. That's important, but it isn't the most important.

As the leader, your team is looking to you for answers. And I don't just mean answers on where to store the mic stands at the end of the weekend services. While a leader must provide for the practical needs of the team, there is a bigger question that needs to be answered: Why?

Why do we mix a certain way? Why do we all have to be here for rehearsal on Thursday night? Why do we only use certain fonts for our worship graphics?

By themselves, these questions are fairly easy to answer. You probably know without thinking twice. The new challenge is that what you do without thinking now needs to be communicated to your whole team. You can't assume everyone thinks like you, and the only way they are going to know what choices to make is if you are talking about them often.

When my children were quite a bit younger, my wife and I took a parenting class on how to help children obey their parents. The main thing I remember about this class was that kids need to understand why they can't do something and not just to be told no. They want to understand the context of the instruction. "Stop running in the middle of the street!" is good advice, but my kids would have an easier time obeying me if they understood that they could be flattened by a car. Rules without context or underlying values are almost impossible to obey.

You aren't necessarily trying to make your production team eat their vegetables, but we do want everyone to be on the same page when it comes to a successful production at our churches. We can't all be on the same page unless someone explains the why behind the choices we make.

Figure Out the How

When I made the switch from doing to leading, I had the luxury of not thinking very hard about why I did certain things and not others. I just did them. Now, as a leader, a big part of the job is to communicate values for the team to follow. For all of us in this situation, we need to have a firm grasp on why we make certain choices over others.

When I was first mixing audio, I remember people asking how I did certain things, or how I made it sound that way. I had no idea, and I usually told them that. As a leader, this doesn't work. You not only have to know what you're doing, but you also need to know the how and why you're doing it so you can communicate it to someone else.

For you to be able to make the jump from doing to leading, you need to sit down and come up with all the things that are important to you and why. It isn't enough just to do things a certain way or to expect your team to be able to read your mind. And unless you can articulate what matters, your team won't be able to follow.

I'm not talking about a list of rules but of values: guiding principles that will help your team frame decisions they need to make along the way to achieve the results your church needs. If all you come up with is a list of do's and don'ts, the people on your team will have a difficult time thinking for themselves. And when things get crazy, which they do in live production, you need everyone thinking similar thoughts on how to get something done and not just following a set of rules that may not apply to that exact situation.

Unless you can articulate what matters, your team won't be able to follow.

Depending on the size of your team, the best way to accomplish the most is if everyone has some level of ownership on the decisions they make. When people are just following rules without thinking, after a while they will engage less and less with what really needs to get done and why.

I love history and I've been watching the mini-series "Band of Brothers" (again). I've lost track of how many times I've seen it now. While it can be hard to watch, for the most part, it makes me proud of the kind of people who serve our country. They aren't fighting for more territory but for an ideal: the freedom of people.

One of the biggest distinctions between the way the American military operated in that conflict and the way their opponents operated was that they expected commanders in the field to think for themselves—to improvise, depending on the needs of the moment. In the other armies, local commanders were tied to the overall plan with no option to alter it. As a result, the American forces were able to use the values that were drilled into them to make decisions in the moment.

When you teach your team the values of production ... and not just production values, but values on how your team does production specific to your environment, it is able to participate more, which means the weight of responsibility is spread across the whole team, not just on you, the leader.

Figure Out the Why

Many production values aren't rocket science. Almost any well-done production follows similar values. So, what makes production in the local church different? What separates what we do from what happens at an average concert?

There needs to be something bigger than just doing production with excellence. There needs to be more than just being as prepared as possible so you can be as flexible as possible. What we get to do in the local church has to be about more than just doing production well.

You're looking for some kind of statement that gives you and your team an ideal to shoot for, something that is larger than simply getting the task done well. Doing the task well is something you aspire to, and it might require a stretch.

I've been listening to Dave Ramsey's book "Entreleadership." When he talks about his company's mission statement, which is something like "To empower and bring hope to everyone from the financially secure to the financially distressed," he says every task and every person's job can be explained by how they bring hope to people. Whether you answer phones, ship products or plan events, everyone is working toward bringing hope to people. When you can describe your job in terms of a higher purpose, even the most mundane tasks have a greater meaning.

What we get to do in the local church has to be about more than just doing production well.

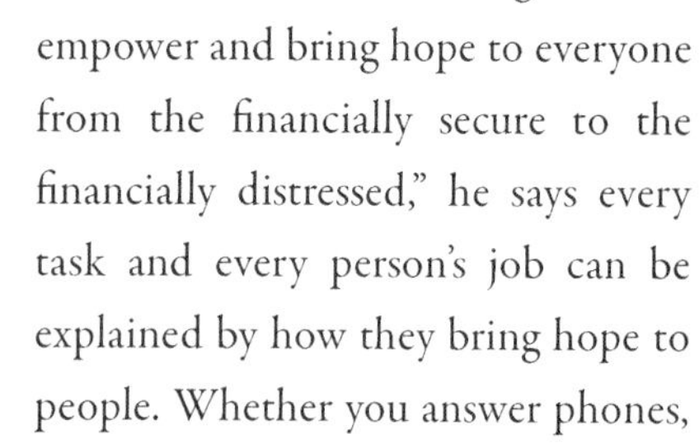

At Willow Creek, our production team mission statement was "To create life-changing moments through the fusion of the technical and creative arts." Whether you are wrapping cables or editing a video, this statement reminded us that we are trying to create a space for people's lives to be changed by God. It also let us know that we are on a team with creative artists to create something together. Without great content, the best production in the world can't create life-changing moments. We need to work together in order to achieve the mission.

When you can describe your job in terms of a higher purpose, even the most mundane tasks have a greater meaning.

Repeat Often

Mission and values can't be something you talk about once and expect everyone to get. As a leader, this can be tough to come to grips with. People will start to get what you're saying at about the same time you start feeling like a broken record.

There was a season when every time the team got together I would ask who knew our mission statement. I would usually offer a gift card of some kind to make it interesting. Typically, there was a pause while people tried to remember what it was. Almost every time (OK, every time), the same person would raise their hand and recite it.

I kind of got tired of asking the team, and so I gradually stopped bringing it up. Over time, even I had to think pretty hard about what the mission statement was. That's pretty sad considering I'm the one who wrote it in the first place!

When you start thinking about the volunteers on your team who serve only once a month, you realize they have even less opportunity to hear what the mission and values are. So even if you talked about it every weekend, you'd need to give the same talk four times in a row for everyone to hear it once!

Some of the hard work of being a leader is doing something that might feel monotonous. It is similar to the mundane tasks that must be done in production: taping cables down, watching the whole video before showing it in the service, cleaning the stage after an event. I'm not sure anyone loves to do these things. It isn't why we got into production in the first place. But all those tasks can be traced back to the mission statement. In order to create life-changing moments, these things need to be done.

In Horst Schulze's book called "Excellence Wins," he reviews the 26 value statements he created for the Ritz Carlton Hotel Chain when he was the CEO. Every shift in every department starts by going over one of the values for about 10 minutes. Every Ritz Carlton hotel around the world covers the same values on the same day, and in the course of a month, they cover all the values at least once. What a great way to remind the teams of what really matters!

Figuring out your mission and values needs to happen so the whole team can be on the same page, working toward the same end goals. Then take the time to communicate the mission and values to the team so everyone thinks about production in a similar way. Everyone should then be able to put their specific task into the context of what matters to the whole team and what the team is about.

Chapter 34 Discussion Questions:

1. Do you know why certain things matter to you, or why you do things a certain way? Take time and start figuring out how to communicate these values to your team.

2. What makes your production ministry different from the production happening down the street at the local concert venue? Write a list of words that describe this difference, then try to distill it down to a single, memorable phrase.

3. When was the last time you talked to your team about what mattered? If you have a mission statement, could any of your volunteers recite it?

4. What are some creative ways you could communicate mission and values in order to get your team on the same page?

35

LEADING VOLUNTEERS

I had my first real full-time job during one of my summers home from college. It was a great experience, mostly because I learned I didn't really want a job like that when I graduated. My favorite part of that summer was how I got to spend my time after work. This was the summer before Kensington Church officially launched, and there was a ton of work to be done. I only did two things that summer ... working at a job that paid well but didn't capture my passion and volunteering at Kensington to help get things up and running. I didn't get paid and loved every second.

These are some of my best memories, and they were a determining factor in the direction of my life. So, when I think about inviting people to volunteer on the production team, I try to remember this moment and how influential the opportunity to volunteer was on the trajectory of my life. Not everyone who volunteers will want to become a full time TD in their local church, but there are opportunities for people to really experience the body of Christ through serving.

God's Design

The Apostle Paul talks a lot about the functioning of the body of Christ in 1 Corinthians 12 and 13—some of my favorite parts of the Scriptures (if you haven't figured that out by now). Throughout much of chapter 12, Paul talks about all the different kinds of gifts that people have and how they are used to help the Church function. At its most basic level, building a production team around volunteers is what the body of Christ is all about: people using their gifts so that the church can function.

It is easy to think we are asking people to help us with the tasks and that people are doing us a favor by serving. This is a very human way of thinking, and I confess to falling into this mindset more often than not. If you're a leader of a production team, it is key to have this foundational principle locked into your brain. With the wrong motivation for leading volunteers, we can end up working with volunteers who are serving out of guilt, or for those of us leading the volunteers, leading out of guilt.

Leading volunteers starts from a place of understanding that we are all designed to play a part, and it is up to us, as leaders, to set the stage for people to serve from this place.

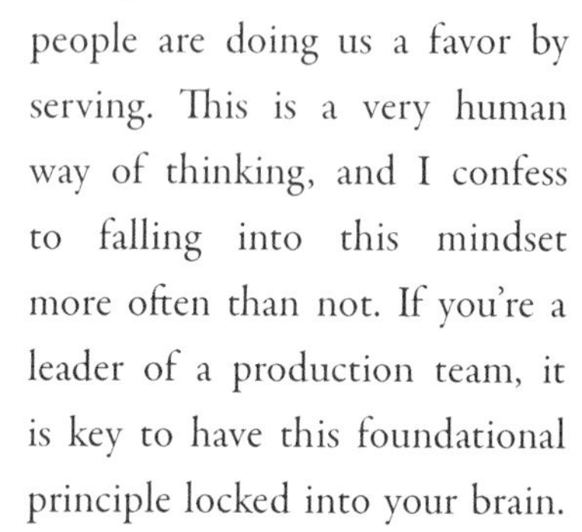

God has created each of us to be a participating member of the local church and not just a bystander, observing from the sidelines. Leading volunteers starts from a place of understanding that we are all designed to play a part, and it is up to us, as leaders, to set the stage for people to serve from this place.

Opportunities to Succeed

Now that you've got a framework for including volunteers on your team, it is up to you to provide places for them to succeed. People often have an idea of what role they would like as a volunteer. Sometimes that works out and sometimes not so much. As the leader, it is your job to create opportunities for people in your church to succeed while serving.

At Willow Creek there are dozens of positions that need to be filled on any given weekend. Some of them are more involved than others. Some require a higher level of skill than others. Not everyone is cut out to run ProPresenter. When I was leading the team there, our goal was to create opportunities for people to do great work. This means that as leaders we need to create these opportunities for people based on their level of ability.

I'm into European fútbol. (In the US, we call it soccer.) It's impressive how many opportunities the clubs provide for players to develop. Their first team is not the only place for people to get real world experience. You've got 2nd team, Under 19s, Under 17s, etc. Not only do clubs have league play, but they have other competitions where they can substitute their best players for those who need time in the hot seat.

When I was at Willow Creek there were nine venues that needed some form of technical support. Some of the rooms were for two-year-olds, with very low-risk production needs. Others were for adults that required a higher production value. And every type of production in between. This wide range of opportunities gave volunteers of all skill levels somewhere to serve.

Expanding the serving opportunities might help start people off in lower risk situations. It might also mean that someone never quite makes it to mixing audio because they wouldn't have a chance to really do well. We should be looking for ways to stretch people and test their abilities in situations that don't carry too much pressure. But it always takes some level of intensity to learn and grow as a technical artist.

Whether you have positions on your team that offer safe opportunities for people to thrive or you have multiple venues for people to learn and grow into positions of more and greater responsibility, it is up to leaders to create experiences where excellence is happening and people are thriving.

Hold Them to the Standard

I've lost count of how many times I've heard the question or wondered myself, "How much I can expect from volunteers?" After all, they're volunteering. I can't *really* tell them what to do.

Think of this in terms of a sports team. I was never a huge participant in sports, but I've played on a few teams. Some of them were horrible, and some were amazing. Any day of the week, I would rather be on the amazing team. I want to win. Who doesn't? The reality is that winning is hard work. It takes discipline. It takes long hours. Doing something really well comes with a price.

If you hesitate to hold your camera operators to the standards of headroom and focus, the camera operators who do hold the standard are going to start wondering if all their hard work really matters, and they will soon become disillusioned with serving. It is important to have worked out the values for your team in order to hold people to the standard. If there is no standard, it is difficult to hold people to it.

When you start asking people to be involved on the team, it is important to be very clear about the expectations. Expectations of time and commitment. Expectations of effort and results. If the people on your team have no idea what the goal is, they are just going to give it their best guess, which, chances are, does not meet the standard.

If there is no standard, it is difficult to hold people to it.

I was having a conversation with someone the other day who asked about this idea of whether to hold people to a standard or not. She was hoping that people would pick up on the values just by watching her. While setting an example of how you want people to work is essential, without some set of guidelines to follow, most people won't automatically fall in line with your unspoken expectations.

Protecting Your Team

You've got a team. They're serving in their sweet spot. They know the standards. Now what?

When you have people who are sold out to the vision of the church and to the team, it is difficult for them to peel themselves away from serving because they love it so much.

I used to be one of those people. When things were going great, I loved it! But it wasn't long before I was frying myself. Nobody at the church was asking me to dial it back, and I fell into thinking that if I didn't do it, it probably wouldn't get done. I just had to tighten my belt and get to it.

Thank God for my wife, who even this morning said, "Come home early!" It was her way of saying "Enough." She has helped me set boundaries for when and how often I served.

While I don't want to say no for my volunteers, it is important that we are helping set realistic expectations on how often people on our team serve. From a developmental perspective, if you're always using your best people in all the positions, there is no room left for others to learn and grow into those areas. Pretty soon you have a team of tired superstars not performing at their best—and no bench at all.

One of the best ways to protect your volunteers from overserving is for you to spend more time developing others—to train-up replacements for your star volunteers. For me, problem solving and getting ready for the weekend was the thing I spent most of my time on. It was way easier to deal with a task list than work with people. Recruiting and developing volunteers ended up last on my list. "I can do this faster myself" meant that nobody else was learning how to make our services happen.

Help your volunteers go the long haul by building a team with a deeper bench.

Chapter 35 Discussion Questions:

1. Do you approach asking people to volunteer as a favor to you or as an opportunity to engage in the body of Christ?

2. Are the volunteer opportunities designed for people to succeed? How many entry-level positions do you have for people to get started on your team?

3. Do you have standards that your team knows they are trying to hit? If not, make some. If so, how often do you talk about them?

36

HANDLING MISTAKES

Remember the lighting console failure right before the start of the Willow Creek Global Leadership Summit? We had what's called a "brown out," which is basically a dip in the electrical supply. It wasn't enough to kick us to UPS power but enough to crash some of our lighting equipment. That's when every production person's "favorite" moment happened. People in the audience turned around in their seats to glare at the booth, wondering when things would be fixed. Meanwhile, I was in the front row wondering when I should start to panic.

For those of you who know me, you know panic isn't my thing. (Wait five minutes, then freak out!) I realized that we had the smartest people in the building working on the problem, and we'd be functioning as soon as possible. In the time it takes Windows to reboot, we were off and running again.

Mistakes happen. I'm reminded of that every week. The reality is that stuff happens—stuff that we can't plan for. In my opinion, there are two kinds of mistakes: ones that happen because I didn't prepare well, and ones that are out of my control. The latter just happen. There is no way to stop these kinds of mistakes, but there are ways for us to manage them—and ourselves—when they do.

Let it Go or Take Control?

When you work with volunteers each week, it can become easy to do most of the difficult stuff ourselves. After all, we know how to do this stuff so we could do it much faster. In the short term, this saves us time each week. In the long term, we are spending a lot of time on stuff that other people could learn to do and do well.

The trouble with planning for the long-term is that it requires living through a certain amount of immediate pain. For someone to learn how to do something requires them to live through all that comes with it: the obvious parts, the parts that are easy to forget about and the crisis that can happen in the midst of it.

For people to feel ownership and to feel like they are not being micro-managed, they need to be responsible for all of the above—the good and the bad.

But as a leader, it is important to not give people too much ownership if they can't handle it. Responsibility is something that needs to be released over time in ways that offer the greatest chance for the person to succeed at the task.

OK, so let's say that we have given someone appropriate amounts of responsibility, and something bad happens in the moment. How do we decide when to jump in and when to let the person figure out what is happening?

First, I wait a few seconds, allowing the other person time to react. Seconds seem like hours, but I try to remember that it is only seconds.

During these few seconds, I try to measure how big the mistake actually is. Will the service come to a standstill? Will only a few people notice? Could I solve the problem faster than the person in the seat?

Depending on the answer to questions like these, I will do one of three things: let the person figure it out, offer a verbal suggestion, or jump in myself. Each response will be in direct correlation to what I know of the situation and the person in the situation. If I let them work through it, will it tank the service? Will jumping in save the service?

There is another layer to this idea of mistakes. Is this the first time someone on the team has made this mistake or is this an ongoing problem?

First-Time Mistakes

Can you remember what it was like to be sitting behind the audio console the first time something bad happened? What if you could go back and look at the first few videos you edited? Are they perfect?

In our world of church production, very little gets accomplished without volunteers, and in most situations, the volunteers involved aren't professional production people. They are middle school kids, or accountants or empty nesters. They may not have the skills right out of the chute, but they have a heart to serve their church, and they've chosen production as their ministry. With so many non-professional technicians, we are looking at a pretty intense learning curve.

If your church is growing, the production needs of your church are growing, which means your team needs to grow.

With any type of learning, people are figuring out what works and what doesn't; they are doing the best they can in each moment. Because they don't have much experience, they don't know where the land mines are, and they can't predict how something will turn out, because they've never done it before.

The challenge of a leader is remembering what it is like to be learning something new. It was a lot easier for me to handle mistakes when I didn't know any better, and we were all learning together. Once I had learned from a particular mistake, it was easy to expect that it won't happen again.

In any team situation, you never know when turnover is going to happen, so you always need to be developing new people. If your church is growing, the production needs of your church are growing, which means your team needs to grow. People need to be developed to handle the increasing needs of your church.

This kind of environment requires a culture of learning. A culture of learning needs room to make mistakes. If there is zero tolerance for mistakes, how can we expect people to learn? First-time mistakes are essential for each volunteer's development as a team player. I'm not saying that we should throw a new volunteer behind the mixer in the main service and call that success. Where the expectation is for people development and not just excellence above all else, there should be opportunities for people to learn how to mix.

We shouldn't try to make mistakes, but we should also have a level of tolerance for mistakes that allows our volunteers to get better through the process. If we don't allow space for first time mistakes, we are saying that we want everyone to be perfect right away. That creates an impossible situation and severely narrows the number of people who can serve at your church. This isn't realistic.

I love first-time mistakes, which implies I'm probably sick in some way. First-time mistakes mean that we're trying new things and stretching ourselves. But these first mistakes have to be addressed or else we'll miss an opportunity to learn from them. Don't just let the mistake go or make excuses for it. Leverage each first-time mistake, by either a new volunteer or an experienced veteran, as a development opportunity.

Values Check

I make sure I talk through the mistake with the team, figure out how/why it happened and how to make sure it doesn't happen again. Critical to following up a mistake is to make sure that your team understands why it was bad, and what values the mistake might have violated. Without this step, it doesn't matter how you respond in the moment. You need to have a mechanism for correcting the mistake.

This goes back to the idea of writing down things that matter to your team. If there are no expectations set before a mistake happens, it is difficult to figure out why a mistake is bad.

Celebrate those first-time mistakes. They can reveal values that are missing from our list, and they have the potential to make the whole team better. If we skip this step and don't analyze our mistakes against what matters to us, that first-time mistake could become a second-time mistake, which is a problem.

Second-Time Mistakes

I was the production manager for a conference, and I was developing someone to call cues. He had done it before but in smaller settings under less pressure. I knew things wouldn't be perfect, but I felt it was a risk worth taking.

As we got into the first session, he made the mistake of not cueing the Front-of-House engineer, and so we missed the beginning of a speaker's talk. After things settled down, I leaned in and we talked about it to make sure he remembered to do it right the next time.

Then it happened again in the next session. I leaned in and reminded him.

Later that day, it happened for the third time. This time the senior pastor was in the booth looking over our shoulders. Great! I leaned in and reminded him again.

At the end of the day, this person asked me how he did and if there was anything that he could improve on. I said, "Let's stop making the same mistake over and over again. I'm ready for you to make a new mistake!"

Making a mistake the first time is all about learning. Making a mistake a second time is careless and lazy. And if we all know what our values are, why does that person keep making the same mistakes?

Second-time mistakes are when being a leader of a team really gets fun. You will need to address this with your person. It is either a problem of not communicating the values often enough, or not communicating the values clearly, and no one really knows what a mistake is.

Maybe it means this person is serving in the wrong area, and they're making mistakes because they're not well-suited to the task at hand.

Whatever the reason for the second-time mistake, it comes down to a leadership issue.

You're the Leader. Take Responsibility.

If mistakes are happening in a service, the number-one thing I do as a leader is to take the heat for the mistake myself. I don't throw the team under the bus; I own the mistake for my team. My church's leadership is looking to me for answers on why mistakes are happening. Saying, "So-and-so really blew that transition" isn't helpful. Not only is it a horrible way to lead your team, but your leaders are wondering why you're in charge of a team if you're blaming them. Why are they serving in that way if they aren't capable?

The other lesson I have learned is that most times a production mistake is noticed by everyone, including the people who lead you. Don't wait for them to address the mistake, bring it up first. This acknowledges to the leader that you know it was a mistake and that you will do something about it. If you wait for your leader to address it, she might be wondering if you even think a mistake happened. Be proactive.

Generally speaking, mistakes can be linked to a process issue rather than a people issue. I always blame a process before telling my leadership that a team member is bad. The process could be cue-sheet related; it could be that person wasn't ready to sit in the seat (my decision) or a step was missing in getting ready for the service.

Above All, Relax

When I was a kid, my uncle would take me golfing. When it was time for my first drive, he would say something like "Bend your knees. Keep your left arm straight. Keep your eye on the ball. Visualize yourself hitting the ball." This list went on and on. Then at the end of it, he would say, "Above all, relax!"

You can't spend enough money to make mistakes disappear forever.

Thanks a lot! I have so many things to think about, how can I possibly relax?

I think there is some wisdom here that is relevant to what we do week after week.

It is easy to think that a mistake is the end of the world, but worse things have happened.

It is great that we care deeply about creating distraction-free environments, allowing people to experience God without production getting in the way. However, it can be easy to take ourselves too seriously.

Work hard. Cover your bases. Respond to mistakes. Figure out how to make sure they won't happen again. Then get over it. Move on.

Our job as technical artists is to make sure mistakes don't happen, but as flawed humans, we must realize that mistakes will happen. The unforeseen happens. The unplanned-for happens. You can't spend enough money to make mistakes disappear forever.

Push yourself to do your very best to eliminate mistakes. Then let it go. If you are slacking and mistakes are happening, that is one thing; get it together. But if you are doing your best and if you are practicing excellence, then give yourself a break.

Chapter 36 Discussion Questions:

1. Are you giving your team the space to make first-time mistakes? Or are mistakes forbidden?

2. How do you represent mistakes to your leadership? Do you blame people or process?

3. When was the last time you shared your production values with your team? How could you bring the values to the forefront of your team members' minds?

37

CELEBRATE THE WINS

Much of your team's culture is created in the heat of the moment. When things are going great or going down the tubes during a live performance, how are you dealing with these situations? Not only how you are treating your people (staff or volunteers), but how are you personally handling all the moments that make up a live event, good and bad?

Fear of Failure

I have worked in a few environments where most of the crew was afraid to screw up. We used to jokingly say to each other "Don't screw up!" before we started any big production. While we were laughing when we said it, it was no joke. Most of us were motivated by fear.

As a result, some of the most talented technical artists I knew were doing their best out of the fear of failure. I don't know about you, but I would much rather do my best out of the love of doing great work.

I was the technical director for a service where the audio engineer forgot to turn up the string quartet we had on stage. I leaned over and whispered "Quartet." He did a facepalm and turned them up. After a few seconds went by, I leaned over and whispered, "Lead vocal." Another facepalm.

After the fact, the audio engineer thanked me for not losing it on him but calmly telling him what was happening. I remember thinking that I wanted to be yelling because we were missing it, but I knew that yelling would probably only fluster the engineer. A flustered engineer is potentially going to make more mistakes. I'm good with the few we've already made.

I also believe that people want to do great work. Believing this is a much more useful starting point for leading teams. People are already afraid of making a mistake without me making it all about not making any mistakes.

Part of what can lead to this fear of failure is the fact that, generally speaking, we only hear about and talk about the mistakes we've been making. We are motivated to always make stuff better. The non-tech people in the room want to make sure we noticed that something wasn't right. (Thanks a lot!)

Celebrate the Wins

When people are doing a great job: following the worship leader doing a song that wasn't planned, lighting that matched the moment exactly, the perfect mix (not possible, but you get the idea); do you let the members on your team know they are killing it?

Sometimes I like to respond in the moment with a "Great job" but not at the expense of distracting that person from the task at hand.

I've noticed that often we spent lots of time talking about wasn't working, and we never really talked about all the great work people were doing. I decided our team needed a "Time of Affirmation" at the end of our Saturday-night-service debrief. The goal was to talk about the good that was happening.

There are a couple things I love about the Time of Affirmation.

If we don't encourage each other, nobody will ... or they would have been doing it already.

Since I knew I would need to share some type of affirmation at the meeting later, I was paying more attention to all the great work happening during the service. Instead of just seeing all the mistakes, I was also looking for the good during the event.

The other thing I love about public affirmation is that people get affirmed publicly! Write that down!

I don't do my job so that people say, "Way to go!" But it is great to hear, and it motivates me to keep doing my best. The people on your team need to know that someone noticed their attention to detail, or their hustle at a particular moment or that the band's monitor mix was like bathing their ears in champagne.

If we do our jobs properly, normal people won't notice. The average attendee in the congregation sees the things that go wrong. Mistakes are easy to point out.

So, when we are gathered as a production team, we are the ones who know about all the ways we crushed it, but too often we don't say anything. Why not take advantage of encouraging each other with a "You crushed that transition!" or "The way you set that flip chart was amazing!" And always, "The best monitor mix ever!"

If we don't encourage each other, nobody will ... or they would have been doing it already.

Chapter 37 Discussion Questions:

1. Is there an atmosphere of fear surrounding your production team? What is one way you could lessen that?

2. Think of someone on your team who does great work. When was the last time you affirmed them?

3. How can you build affirmation into how you do production at your church?

38

TASK v. COMMUNITY

As a group of technical artists, the whole reason we're together is to get a task done so our churches can function.

The challenge with being so task-driven is that it can be easy to start seeing the people you're leading as a means to an end and not as real people with real gifts to help build up the body of Christ.

On the surface, this focus on the task can make it seem like the task is the most important thing. In reality, people are the most important thing. The people we work with, the people who are on our teams, the people who make it all happen, the people in our congregations.

If all we focus on is the task, our people will start to feel like they are only good for getting a task done.

When I first started volunteering in production, I did it for the cool stuff I got to get my hands on. The EV Tapco 100M-powered sound board. That sweet TEAC cassette deck. The AKG D-109 Lavalier microphone with the super handy necklace to hold the mic in place. I was living the dream.

After a while, the gear became less and less important. What kept me coming back, besides my somewhat over-inflated sense of responsibility, were the people. I was becoming a part of something. I belonged. The other people who were there early on Sunday morning were becoming my people.

The challenge with being so task-driven is that it can be easy to start seeing the people you're leading as a means to an end and not as real people with real gifts to help build up the body of Christ.

Later in my journey as a technical artist, there was

a time when I felt like I was only wanted for the task I could perform. As a volunteer, that feels pretty icky. Maybe this wasn't true, but I felt icky. I stopped serving.

As leaders we can be obsessed (for good reason) with the task at hand. However, when you lead a team, it is just as important to care about the people. The team culture you create as the leader will determine how much can be accomplished.

Make Time While You're Together

I generally thought the only way to build a team culture was to add more events to our team's calendar. Get everyone into a small group. Have a team-building exercise once a month.

But if your teams are serving all weekend long, how can you be more intentional with the time you are already together? Create space in the schedule that doesn't involve a task. Share a meal together. Meet 10 minutes early to circle up and share a scripture and the plan for the day. Maybe even ask for prayer requests.

Make Time to Be Together

Creating space in the times when we are already together is a key way of building community. Adding a little margin to the schedule won't generally break anyone's ability to commit, especially when you are intentionally making it worth the extra time commitment.

Outside of the times we were already together, my wife and I would occasionally schedule a "Spaghetti Saturday" for the team. After our Saturday night services, we would invite people to our house, families and all. There was no agenda other than to be together and relate to each other on a level beyond the task.

An event like this was a lot of work, especially for my wife since I was at work while she was trying to get the house together. And my oldest son won't eat spaghetti anymore. But hey, taking one for the team!

Volunteers can start serving because of the gear, but they keep coming back week after week because of the relationships, which points to the need to invest in our people. Sure, we need to be good at the tasks, but we also need to care for each other.

My friend John Cassetto, the Global Worship Pastor at Saddleback church, says "People on our teams shouldn't *feel* like they are loved, they should actually *be* loved." In order for this to happen, we all need to focus more of our attention on how to love the people on our teams.

As a tech person, knowing what the task list looks like is the easy part. However, being intentional with the relationships before us is the key to long-lasting, meaningful relationships.

Chapter 38 Discussion Questions:

1. Think about your own production journey. What are some of your most memorable moments? Were they about the task or about the people?

2. If you have multiple services, what are you doing during the in-between times? What are a few things you could plan to do with your team to help build community?

3. What community-building activity could you make a regular part of your team's routine?

39

YOU SET THE TONE

As the leader, people are looking to you. If you're anything like me, when I started moving from doing the task to leading the people who are doing the task, I was pretty uncomfortable. I like being in the back of the room. I chose the production life to be in the background, people! So, stop looking at me! If you are the leader, people are watching you and will follow your lead. Good or bad.

Without Grumbling or Complaining

How often does your leadership have an idea that seems idiotic to you? How often have you shared your frustration with the rest of your team?

Shut up. This is a tough one.

Being a leader is difficult, and it's never more difficult than in a situation like this.

Your whole team is complaining, and you want to join in. But if the church is going to be as effective as it can be, you are going to need to point your team in a more positive direction.

Philippians 2:14-16 lays it out there for us:

> *Do everything without grumbling or arguing, so that you may become blameless and pure, "children of God without fault in a warped and crooked generation." Then you will shine among them like stars in the sky as you hold firmly to the word of life. And then*

I will be able to boast on the day of Christ that I did not run or labor in vain.

I'll say it again. Stop complaining. Just shut up!

Your team is always watching. Not only watching to see how production needs to be done at your church, but watching how you respond in each moment. Will you join in the grumbling, or will you rise above it? Will you add fuel to the fire, or will you put out the fire?

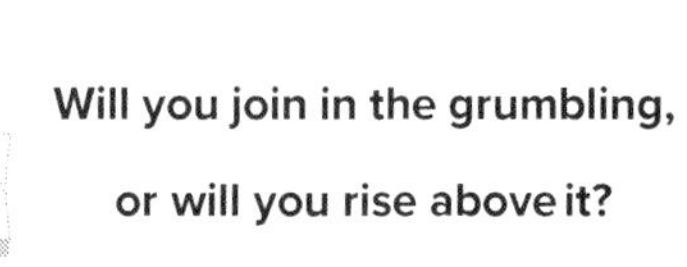

If we look at the natural state of things, we will always be descending into chaos. Physics tells us so in the 2nd law of thermodynamics:

An isolated system will gradually decline into disorder.

Our team's system requires that we are investing energy to keep it moving forward. Teams won't have great attitudes about the changes coming at them unless you, as the leader, are pushing things in a positive direction.

An isolated system will gradually decline into disorder.

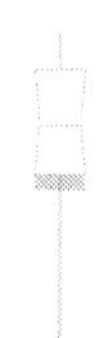

You're the Leader, Now Lead

As a leader, one of the more difficult tasks is inspiring the team to live out the team's values week in and week out. Heck, it probably really starts with inspiring yourself week after week! It is the leader's job to call the team to something bigger than the task at hand. Why are we doing the task in the first place?

Once you've figured out what your mission and values are, post them where people can see them; have a contest to see who can remember them all; every time you start a load-in, talk about what matters. The way to get your team on the same page is to talk about these things all the time.

Whether you know it or not, you are creating a team culture. You are either doing it on purpose or it is just happening. If you aren't being strategic with how your team functions, they will be functioning however each person feels like they should.

If you want the people in the booth to greet the band when they arrive, you need to talk about it, and they need to see you doing it.

If you want everyone to bring their A Game to the rehearsal, have you told the team that? Do they know what their A Game is? Are you bringing your A Game?

The team is watching and listening. Are you running? Are you grumbling? Or are you pushing your team in a positive direction?

Chapter 39 Discussion Questions:

1. Considering how your team functions, what do you think matters most to your team?

2. Think about what you'd like your team to be. Write down five characteristics to describe this future team.

3. What are some creative ways you could communicate team values to your people?

40

LEAD YOURSELF

When I watch a high-capacity leader, it generally seems like they have it all together, and stuff comes naturally to them. When a person has mastered something, they make it look easy. The reality is that there is work going on behind the scenes I never see. From the outside, I only see the tip of the iceberg of what goes into making something a reality.

If we hope to lead a production team that is thriving, we need to be healthy ourselves. We need to lead ourselves to become the best possible version of ourselves. I don't think it is an overstatement to say that many of us work, work, work and think about our own health last.

We all want our teams to do amazing work. For that to happen they need to be led well. For a team to be led well, we, as leaders, need to be at our best. Here are a few things I have tended to miss in the quest for becoming the best version of myself.

Go Home

Guess what? You will never get all the work done. It doesn't matter how many hours you work in a week. The pile will always be there. You know what else? There is life outside of your job.

I've worked at a couple of churches where the work expectations were pretty high. We worked long hours. Plus, we were killing it, and that is always a great feeling.

Once I was walking to my office, and I bumped into the senior pastor at around 5 p.m. He said "Don't stay too late!" as he was walking out the door. I told him I was wrapping up and that my wife and I had an agreement that dinner was on the table at 6 p.m., so I needed to be home on time for that. He said, "At my house, dinner is on the table at 5:30, so I've got to get moving!"

This exchange was a huge revelation to me. So much of our work culture seemed like it came from the top of our organization. But here the top guy was saying he had to leave at a reasonable hour to make it home for family dinner! If he can go home for dinner, I shouldn't feel so guilty about doing it myself.

Much of my need to stay later than everyone else was so I didn't have to confront my own dysfunctional attitude that I've not done enough.

If I'm the last one out, then I can feel better about myself and my contribution. If you're familiar with the Enneagram, I am a classic Type 9—the Peacemaker. What I should be doing is dealing with my wrong thinking and being OK with leaving, even when others might still be working.

As I mentioned earlier, when I would leave the house in the morning, my wife would say, "Come home early!" My response was normally nervous laughter. Did she have any idea how much work I had to get done? Maybe she meant early in the morning. Like 1 a.m.?

While I wasn't able to come home early all the time, it was usually on my mind while I was at work: how can I work efficiently so I can leave early? And especially now that it has become part of my everyday life, I can't imagine life without that simple phrase early on in my journey.

Life exists outside of church. Go live it. It makes you a better human.

Read Your Bible

It sounds like a joke. Of course I will read my Bible. I'm a Christian, right?

Actually, it isn't a joke.

For those of us working in the local church, we are there all the time. But how often are you getting fed by being in church? If you're working, it is difficult to pay attention to the message like normal people. It is difficult to worship the same way others do. If we aren't careful, we will empty ourselves out for the sake of the church and be left with an empty shell.

In reality, it isn't the church's job to feed you spiritually. It is your job. I think many Christians miss this truth. We all look to our pastor and the church staff to do our spiritual work for us. I'm not saying that God can't move and work in a church service. That's kind of the whole point. But in truth, God wants a personal relationship with you, which can only happen if you are spending time with Him alone.

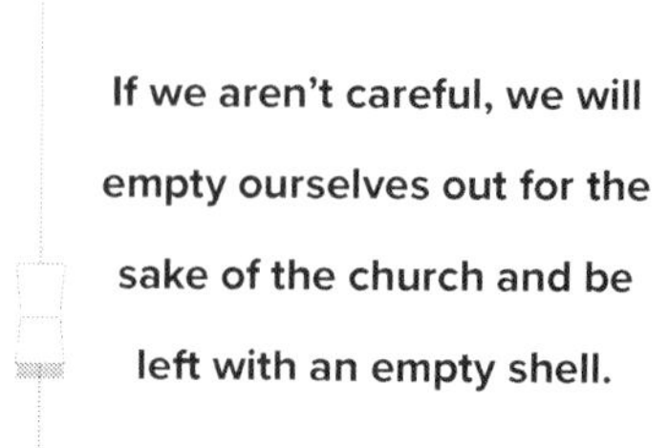

There was a time in my life when I decided I would start reading my Bible every day. Don't all good Christians already do this? I told friend, "I know people tell us all the time that we should read our Bible, but it actually works!" I became more attuned to what God was doing around me. I became more acquainted with my depraved heart and more dependent on Jesus.

We need to develop the disciplined, consistent habit of spending a few minutes each day absorbing the words of God. If we want to become the best version of ourselves, we can't do it without filling our brains with the truths of Scripture.

I became more acquainted with my depraved heart and more dependent on Jesus.

Get a Hobby

For much of my life, I only did three things: work, go home, eat, sleep. If you do that for long enough without much variation, you can dry up pretty fast.

As I said at the beginning of this book, we are all artists. What we do requires a level of creativity. If you don't take time to refresh yourself, you will empty yourself out.

The beauty of developing a hobby is that is gives you something to look forward to. It also has a way of distracting your brain from the overwhelming problems to be solved at church or home. And if you've signed up for a class, it gives you a reason to leave work at a reasonable time.

Maybe it is a home renovation project. Maybe it is a cooking class with your spouse. Maybe it is stamp collecting. The exact hobby matters less than the fact that you're involved in something completely different from what you do every day. This will give your brain a rest from the problems at work. By expanding your interests, you will develop a new perspective on life and work.

Just as our bodies need a chance to recover from exercise, our brains and hearts also need chances to recover: to remember why we love doing production in the local church in the first place, to give our subconscious a chance to figure out some problem that needs to be solved.

Eat Better

I'm a huge fan of production food. You know, stuff you can eat while you work. Pizza. Hamburgers. Chips. Cookies.

I also have to stay awake, so Mountain Dew is a necessity.

It is easy to let ourselves go when it comes to what we eat. If we hope to live healthy lives and have the energy to lead our teams, we need to be taking better care of ourselves.

Corresponding to the idea of your team following your lead, what kinds of food are you providing for your team? Is it whatever is easiest? I totally get it, but there is a better way.

I would argue it is an issue of pace. I think we are usually moving too fast, and we don't value our time like we should. As a result, I'm always thinking about the fastest way to feed people and keep moving. There is nothing wrong with stopping for a few extra minutes to eat something that is healthy and less portable. It will help provide community moments with your team members that you wouldn't get otherwise.

Get Off Your Backside

I know that production work is hard on your body. I wake up sore more days than I care to mention. But for the majority of our lives as technical artists, and really as leaders of technical artists, we spend most of our time in meetings, working on spreadsheets or sitting behind a console for hours on end.

I need to create time to actually move. Many of us are living right at the edge with no margin, and making time for exercise can be difficult. It sounds like a nice idea, but that's about it.

It can be as simple as walking your dog more often. Ride your bike around the block. Park at the back of the lot and walk into Home Depot.

You don't have to sign up for a marathon. You don't have to start a hot yoga class. It doesn't have to be an Ironman competition. But for the sake of your team, your family and your own health, move more.

One of my favorite parts about introducing exercise into my schedule was that it gave me a chance to ponder the problems I was trying to solve. It also helped me relieve any stress I was carrying. I didn't generally listen to music or a podcast; I just ran. It gave my brain the space it needed to explore possible solutions. I made it through several very tough situations at work by running when I got home.

Without some intentionality, our lives will easily get out of control. Forget about your team for a minute. You deserve to take better care of yourself than you probably are doing. In 1 Corinthians 6, Paul says our bodies are the temple of the Holy Spirit. In verse 20 he says, "Therefore honor God with your bodies." How is your life reflecting this reality?

I need to do better in all these areas. I don't want to get to the end of my days and wish that I hadn't let life drag me along behind it. I want my life to reflect the things that matter to me and not just be a shell of a human who let the urgent squeeze out the important.

Chapter 40 Discussion Questions:

1. What is your biggest challenge in the areas of leading yourself? Is it exercise or eating right, spending time in God's Word, or all of the above?

2. What are some simple boundaries you can set to help you go home at a reasonable hour?

3. Life as a technical arts leader can be never-ending and crazy. What are some small ways you can be intentional about the time given to you?

CONCLUSION

First in. Last out. It isn't for everyone.

Thanks for going on this journey with me. Hopefully you've learned ways to become the best technical artist you can be. As a result, I hope that the rest of your team and your church are driven to be more effective by the movement happening in your life.

When we have a better understanding of how our gifts fit into the rest of the church, we realize that we are better together.

When we have wrestled with what really matters from a production standpoint and then executed from those values week after week, our churches transparently communicate the gospel to our communities.

When we do the difficult work of collaboration with our creative peers, we make better and more effective art.

When we lead our production teams to live out all of the above, I believe that it is a picture of how God imagines the church should function.

I've been reading through the book of Colossians and keep coming back to a section in chapter one that captures my heart for those of us on the FILO journey.

COLOSSIANS 1:9–14

For this reason, since the day we heard about you, we have not stopped praying for you. We continually ask God to fill you with the knowledge of his will through all the wisdom and understanding that the Spirit gives, so that you may live a life worthy of the Lord and please him in every way: bearing fruit in every good work, growing in the knowledge of God, being strengthened with all power according to his glorious might so that you may have great endurance and patience, and giving joyful thanks to the Father, who has qualified you to share in the inheritance of his holy people in the kingdom of light. For he has rescued us from the dominion

> *of darkness and brought us into the kingdom of the Son he loves, in whom we have redemption, the forgiveness of sins.*

These verses speak so well about how we should live as technical artists.

"Bearing fruit in every good work."

"Growing in the knowledge of God."

"Being strengthened with all power according to his glorious might so that you may have great endurance and patience."

What an amazing set of promises that Paul is praying over the people of Colossae.

We pray this over you, FILO tribe.

God has amazing plans for each of us—not simply making it through the hard work of church production but really thriving as Christ followers. Christ followers who also happen to work behind the scenes—to make our churches the most effective they can be.

FILO OFFERINGS

The FILO Community exists in many forms, but with a common goal: to help technical artists in the local church become more effective so that their churches can become more effective. Here are some ways that you can engage with FILO to further your journey as a technical artist:

FILO Conference

The FILO Community gathers together a few times a year for skill development, community and inspiration. Some of the best practitioners in technology in the local church teach classes on topics that apply to doing production well in the local church. Bringing technical artists from around the world just to be together and share ideas with each other. Providing an environment where each technical artist can be poured into and inspired to continue serving in their local church.

www.filo.org/events

FILO Podcast

Todd Elliott hosts the FILO Podcast, bringing together people to talk about topics that affect how church services happen. Whether it is a production topic like audio for video, or how to make creative ideas come to life, Todd talks with people who want to help technical artists in the local church get better.

www.filo.org/podcast

FILO Resources

Looking for ways to inspire your production team? Need to learn a new skill to take your church's production to the next level? FILO Resources are a collection of tools you can use to help do just that. Whether it is a main session talk or a breakout session, there are hundreds of resources to help you become more effective.

www.filo.org/filo-resources

FILO Coaching

Leverage the experience of the FILO team to help make your church more effective. One-on-one leadership development, weekend experience assessment and FILO Team Nights are all uniquely crafted to give you and your team help in taking your effectiveness to the next level.

www.filo.org/coaching

Made in the USA
Columbia, SC
26 January 2025

52626536R00115